THE POSTGENOMIC CONDITION

THE Postgenomic CONDITION

ETHICS, JUSTICE, AND KNOWLEDGE AFTER THE GENOME

JENNY REARDON

THE UNIVERSITY OF CHICAGO PRESS

Chicago and London

The University of Chicago Press, Chicago 60637
The University of Chicago Press, Ltd., London

Published 2017

Printed in the United States of America

26 25 24 23 22 21 20 19 18 17 1 2 3 4 5

ISBN-13: 978-0-226-34455-3 (cloth)
ISBN-13: 978-0-226-51045-3 (paper)
ISBN-13: 978-0-226-34519-2 (e-book)
DOI: 10.7208/chicago/9780226345192.001.0001

Library of Congress Cataloging-in-Publication Data

Names: Reardon, Jenny, 1972- author.
Title: The postgenomic condition : ethics, justice, and knowledge after the genome / Jenny Reardon.
Description: Chicago : The University of Chicago Press, 2017. | Includes bibliographical references and index.
Identifiers: LCCN 2017022052 | ISBN 9780226344553 (cloth : alk. paper) | ISBN 9780226510453 (pbk. : alk. paper) | ISBN 9780226345192 (e-book)
Subjects: LCSH: Human genome—Research—United States—History. | Genomics—Moral and ethical aspects. | Genomics—Social aspects.
Classification: LCC QH431 .R297 2017 | DDC 611/.01816—dc23
LC record available at https://lccn.loc.gov/2017022052

♾ This paper meets the requirements of ANSI/NISO Z39.48-1992 (Permanence of Paper).

For my father,

FRANCIS XAVIER REARDON

(1928–2014),

who made clear to me

that thought and struggle

are not optional

CONTENTS

CHAPTER 1
1 The Postgenomic Condition: An Introduction

CHAPTER 2
25 The Information of Life or the Life of Information?

CHAPTER 3
46 Inclusion: Can Genomics Be Antiracist?

CHAPTER 4
70 Who Represents the Human Genome? What Is the Human Genome?

CHAPTER 5
94 Genomics for the People or the Rise of the Machines?

CHAPTER 6
120 Genomics for the 98 Percent?

CHAPTER 7
145 The Genomic Open 2.0: The Public v. The Public

CHAPTER 8
169 Life on Third: Knowledge and Justice after the Genome

203 Epilogue

209 Acknowledgments

215 Notes

271 Bibliography

299 Index

1 THE POSTGENOMIC CONDITION
An Introduction

What I propose, therefore, is very simple: it is nothing more than to think what we are doing.
—Hannah Arendt, *The Human Condition*

That life is complicated may seem a banal expression of the obvious, but it is nonetheless a profound theoretical statement—perhaps the most important theoretical statement of our time.
—Avery F. Gordon, *Ghostly Matters*

"I believe one day in the not-so-distant future, every person on the planet will have their genome sequenced."[1] So Robin Thurston predicted on July 12, 2016, as he assumed the helm of Helix, a San Francisco–based start-up company that aspires to be the "app store" for genomics.[2] Helix is among a handful of powerful players that today seek to convince people to spit into cups in order that a company, a university, or the state might sequence their genome. Apple, Google, and many federal governments around the world today all have plans to recruit millions—even billions—into the genomic age.[3]

Will they succeed? What exactly would an app store for my genome contain? An algorithm to tell me what percent Neanderthal I am, how related I am to anyone else, or what my risk of a disease might be? Would I, could I, pay to play?

Two weeks after Thurston's prediction of a not-too-distant future filled with sequenced human beings, an article in the *Journal of the American Medical Association* indicated that he might face a hard sell. Its authors asked: "What happens when underperforming big ideas in research become entrenched?"[4] Their first example was the idea that a few genetic variants can explain the causes of common diseases. The notion that common genetic variants link to common diseases, the Common Disease–Common Variant (CD-CV) hypothesis, fueled initial optimism about the power of genomics to transform medicine, yet today few such variants have been found.[5]

During the final decades of the last century, national governments, scientists, and entrepreneurs invested tremendous public and private resources into the idea that the human genome sequence contained the blueprint of life, one that could guide humans into a peaceful and prosperous new millennium. In June 2000, then US president Bill Clinton announced from the East Room of the White House the publication of the rough draft of the human genome, what he described as "the most important, most wondrous map ever produced by humankind."[6] Medicine would be transformed. Cancer would be cured.[7] Racial ideologies that had torn families apart and killed millions over the course of the twentieth century would be defeated.[8]

Yet despite these high-profile public pronouncements, genome scientists knew a hard road lay ahead. They might have produced what they and others referred to as "the book of life," but reading it posed difficult challenges.[9] Over three billion nucleotides—adenines (A), guanines (G), cytosines (C) and thymines (T)—make up the human genome sequence. In 2008, when I visited the Wellcome Trust in London, a major funder of the public effort to sequence the human genome known as the Human Genome Project (HGP), I found 118 books on bookshelves, each one thousand pages long. These books offered me the chance to "read" the human genome sequence.

I remember flipping through their pages, bewildered. What struck me most were the stretches of dashes, regions of the human genome too repetitive for sequencing technologies to decipher. These dashes meant as much to me as the alphabet soup on the other pages.

I was not the only one puzzled. As the jubilation surrounding the completion of the HGP in 2003 ended, a question—at once sobering and exciting—moved to the fore: Now that we have "the human genome" sequence, what does it mean? Tremendous feats of biomolecular engineering produced the sequence. However, what was the route between this technological feat and meaningful knowledge that might foster life and human understanding? In the decade after the completion of the HGP, this turn to the question of meaning—the question of the uses, significance, and value of the human genome sequence—marks what I call the *postgenomic condition.*

Despite its greater import, this task of interpretation that came after the completion of the HGP has generated far less attention.[10] Perhaps this is because there are no clear heroes and villains.[11] Unlike the popular accounts of sequencing the human genome, the story of the efforts to interpret it cannot be told as one of scientific giants battling over good versus

Courtesy: Russ London's photograph of the Human Genome in the "Medicine Now" room at the Wellcome Collection in London.

evil. These efforts are about a much wider range of lives, and a far more diverse range of issues that do not resolve along clear lines that demark good from bad, public from private. They reveal less about the psyches of those heralded as the "great men" of science and more about the conditions of contemporary life.[12] Drawing on the trust I have built over the last twenty years—first working in molecular biology laboratories and then chronicling the emergence of human genomics—in the chapters ahead I share stories that bring the reader into these much richer spaces where the meanings and values of genomic data are being forged.

As we will see, they are stories of these times—times of great promise and despair, wealth and deprivation—in which people around the world are raising questions about how to know meaningful life on a data-rich but environmentally depleted, and interconnected-yet-fractured planet. In 2006, home prices in the United States began to plummet, and in 2008 the world experienced a financial crash followed by several years of recession.[13] In 2016, the political infrastructures at the heart of twenty-first-century Euro-American aspirations of peace and free movement experienced ominous blows: the United Kingdom voted on June 23 of that year to leave the European Union; on November 8 the United States elected a president who promised to deport millions and to build a wall along the Mexico and United States border to keep others out.[14] In the wake of these dramatic and largely unanticipated events, many are asking about the value of things that have long been central to conceptions of the good life, at least in the Euro-American West. What is the value of owning a home? Of investments? Of education? Of government itself? And, importantly, who can answer these critical questions? Anxieties in the current moment are generated not only by the fall in value of these things long viewed as central to security and prosperity, but also by growing distrust and discontent with those entrusted with answering these important questions about value and worth.[15]

Some have diagnosed the problem as post-truth politics. While "willful distortion" continues to plague efforts to right the ship of democratic governance, I argue that there are more pervasive problems that are not so easily grasped or rectified.[16] As I finish writing this book on the European continent, daily there are reports of political turmoil and attacks. In the United States, racial tensions are at their highest in years as police shootings of black men tear at the moral and political fabric of the nation.[17] For many, the central concern is not that leaders lie, but that the world itself does not appear to support or respect the lives of too many. This is not a problem of distortion; it is one of constitution.[18] How is this world put together? Who and what gets to live and prosper in it?

It is my contention that in these times as we rightly turn our attention to correcting falsehoods, we must also attend to the problem of deciding which elements of this troubled world-in-need deserve our all-too-limited energies.[19] Which should be matters of our care and concern? Which should become the stuff of the truths that ground our systems of law and governance?[20] There is, I suggest, a growing sense that dominant liberal institutions of the Euro-American West have ignored too much as they have invested vast resources into knowing and caring about a few things—many of which, like genomes, require investments in high-tech sciences.

The Human Genome: A Thing of Value?

The question of the meaning and value of the human genome is but one instantiation of this broader questioning of the capacities of dominant liberal modes of knowing and governing. Although a perhaps unlikely place to begin an exploration of such fundamental issues, questions of value, trust and truth long have been formative in human genetics and genomics. Human geneticists' role in legitimizing eugenic policies of sterilization and immigration restrictions, as well as Nazi invocations of genetic theories during the Holocaust, left many worried that any effort to look to genes for explanations threatened to either end or undermine the value of the lives of too many.[21] For decades, to study human genetics meant living with this legacy, and responding to these concerns.[22] Human geneticists took part in United Nations–sponsored deliberations about the appropriate meanings and uses of their studies.[23] They changed the names of their journals and professorial chairs, purging the words *eugenics* and sometimes *race*.[24] They invested in developing molecular techniques that they believed offered more objective analyses that could protect their research from social bias.[25]

In the wake of these reforms, some geneticists began to reinvest in studies of human genetic differences. Most prominently, in 1991 human population geneticists proposed the Human Genome Diversity Project.[26] The Diversity Project's proposal to collect DNA samples from so-called vanishing indigenous populations immediately sparked widespread concerns.[27] Indigenous rights advocates dubbed it the Vampire Project.

Biological anthropologists worried that the proposed initiative would re-import colonial imaginaries into human biology.[28] Despite decades of efforts to address concerns, their critiques demonstrated that the anxieties surrounding human genetics lay barely below the surface.

The 1994 publication of *The Bell Curve: Intelligence and Class Structure in American Life* further demonstrated the ongoing salience of these concerns. In this *New York Times* best seller, American psychologist Richard J. Herrnstein and American political scientist Charles Murray argued that a "large genetic influence" on IQ was "irrefutable," and that racial differences in IQ were similarly undeniable.[29] The book, labeled "the publishing event of the decade" by the influential US think tank the Brookings Institution, set off widespread concerns that societal investments in human genetics might once again justify public abandonment of racially marked groups.[30]

Worried that these controversies might taint the HGP and sink human genomics before it even began, leaders of the initiative drew clear lines between their effort and analyses of human genetic diversity.[31] The HGP, they argued, sought to sequence one human genome, not the many that the Diversity Project proposed; it aimed to improve the medical treatment of individuals, not to make claims about human populations, as *The Bell Curve* did. However, as the sequencing of the human genome neared completion in the late 1990s, leaders of the HGP at the National Human Genome Research Institute (NHGRI) significantly changed their position. While comparing the human genome sequence to the sequences of other species—such as the mouse and the platypus—might reveal some things about human evolution, the possibility of genomic understandings of human disease necessitated understanding how *human* genomes differ.[32] Thus, even before the Human Genome Project came to an official close, the NHGRI initiated an effort to collect samples from different populations from around the world—what would become known as the International Haplotype Map Project (or HapMap, about which we will learn much more in the pages to come).

Leaders at the NIH knew the HapMap would take them into ethically fraught terrain. Acknowledging this, in an unprecedented move they brought ethicists into the planning of the project at the very beginning.[33] Further, they pursued an overtly positive approach. While they did not deny the potential for new forms of discrimination, they emphasized the liberating potential of their efforts. Far from the totalitarian and eugenic associations of twentieth-century human *genetics*, they promised a new science of human *genomics* that would be at the vanguard of twenty-first-century antiracist democratic societies and sciences. This new science of genomics would demonstrate that race had no biological meaning.[34] It would forge new kinds of relationships with research subjects that emphasized their inclusion and participation in the whole process of research.[35] Researchers would not just collect blood and leave. They would stay and "give people in the communities . . . an opportunity to share with investigators their views on the ethical, social and cultural issues . . . and to provide some input into the way their samples would be collected and described."[36]

The Rise of Genomic Liberalism

Efforts to secure the meaning and value of human genome sequence data through creating a participatory, inclusive, and open genomics—what I call *genomic liberalism*—made for an exciting decade both scientifically and politically. While liberalism is a heterogeneous political tradition with both equalitarian and hierarchical strands, one of its core concerns is that concentrated power corrupts and so government should be limited and power shared with the people.[37] This core commitment to share power guided the efforts of HapMap leaders as well as the many others who over the decade sought to make the human genome meaningful for broad publics. Foregrounding these modern liberal democratic values of inclusion and participatory governance in what previously had been largely a technocratic expert arena led to nothing short of a remarkable turn of events, one that generated much hope for more just forms of science and technology. Linda Avey, co-founder of personal genetics company 23andMe, called for the end of the "feudal system" in which researchers held all the power.[38] George Church of Harvard University foresaw radically opening up the doors of human genetics and molecular biology so that not just scientists, but all people, could take part. In 2005, in an effort to break down what he considered old, overly protectionist policies that limited citizens' participation in genetic research, he launched the Personal Genome Project (PGP).[39] The success of the PGP gave rise to Open Humans, the newly launched initiative that draws on the PGP's "public participatory" approach to "reimagine health discovery as a collaborative effort" between researchers and participants willing to openly share data about themselves.[40]

These and the other projects we will encounter in this book imagined new democratic frontiers for human genomics that inspired many. In his speech celebrating the completion of the first draft of the human genome, then US president Bill Clinton imagined the society he believed human genomics heralded—one in which all individuals are equal, included, and guided by truths revealed by "the greatest age of discovery ever known."[41] Equality, inclusion, Enlightenment. All these are centuries-old enduring goals of modern liberal democratic societies.[42] In the speeches of political leaders and in the books and research proposals of genome scientists the human genome fosters these ideals. It includes us all, and all of us can participate: the data will be in the public domain; it will reveal our common humanity. As Francis Collins, the director of the National Institutes of Health and former director of the Human Genome Project told me in the fall of 2013, "Under the skin, we are all one family."[43]

Genomics became a part of turn-of-the-millennium hopes for justice through data and democracy. From the National Institutes of Health to Silicon Valley to Tahrir Square, in the opening years of the twenty-first century access to data was framed as the antidote to racist ideology and the path to freedom.[44] "I want my genome," announced 23andMe cofounder Anne Wojcicki at TEDMED in 2009.[45] "I want internet," read one placard in Tahrir Square in the spring of 2011.[46] Internet, Egypt, Twitter, genomes, democracy, revolution: these equivalences evoked hope for a better world in which dictators fell and democracy rose.[47]

Freedom's Friction

Yet forging freedom from data in today's unevenly developed network societies presents significant problems.[48] Who has access to the data? Who will benefit? Consider this quip from a participant at a symposium on personal genomics sponsored by my university: given that many people who make their genomes available for analysis are Silicon Valley technology professionals, personal genomics will likely discover the "Apple gene."[49] The message of this anecdote was clear: genomic infrastructures inevitably invite in some more readily than others.[50]

The worry that biotechnology and human genomics will favor the technical elite while excluding many from the benefits and profits of biomedical research remains a persistent fear, one that gained widespread attention through the popular uptake of *The Immortal Life of Henrietta Lacks*.[51] In this *New York Times* best seller, author Rebecca Skloot tells the story of Henrietta Lacks, an African American woman who in the early 1950s received treatment from Johns Hopkins University for an aggressive form of cancer. Researchers removed her tumor cells without her consent or knowledge and used them to create the first viable cell line. These cells, known as HeLa, were perhaps the first biotechnology and became widely useful. Over the decades, universities and companies made "incalculable" sums of money off of them.[52] The Lacks family, by contrast, continued to live in poverty.

While historians and critical race theorists have for years written about aspects of this case, it would be Skloot's popular retelling that sold millions, and brought the story to national prominence.[53] For Skloot and many of her readers, HeLa joined the Tuskegee syphilis experiment as an exemplar of biomedicine's exploitation of the poor and historically disadvantaged. It illustrated how in this post–civil rights era of formal equality, racial injustice continues.[54] It demonstrated how in an era of biocapital—in which biotechnology and capitalism are fundamentally entwined—alienation of human beings from the value that inheres in their own bodies takes new forms.[55]

Even after all the attention the Skloot book brought to these issues, concerns about HeLa once again reemerged when researchers sequenced the cell line and published it without asking the Lacks family for permission.[56] In the view of the many blogosphere commentators, these researchers once again ignored the humanity of Henrietta Lacks and her family in their pursuit of scientific progress.[57]

While genomic data may be invoked to argue that we should move beyond the use of racial categories, the HeLa story and the stories that appear in this book suggest that an antiracial genomics is not the same as an antiracist one.[58] Genomic liberalism might embrace antiracism, but justice still might not be served.[59] How to respond is an emerging challenge, and one that is at the center of this book.

Genomics and Justice? Reimagining Property, Privacy, Persons, and the Social Contract

In the wake of the broad public engagement with Rebecca Skloot's telling of the story of Henrietta Lacks and the creation of the HeLa cell line, discussions have ensued not only about what a just retribution for the Lacks family might be but also about whether individuals have the right to own and control their own cells, including their DNA. Some argue yes.

At a public panel discussion on personal genomics hosted at the University of California, San Francisco, in 2012, an audience member invoked the Lacks story and then concluded: "I think it is outrageous that people are expected to give all their data. And in this case it really is you because it is your DNA sequence, and you're not allowed to get any money."[60] Others disagreed. "I'm not sure the net present value of your genome is high," responded a panelist at the event. He went on to explain that genomes only become valuable in aggregate.[61]

Passion and a sense of justice lie on both sides of the debate. For the audience member, a genome is the unique essence of an individual—"it really is you"—and so, in a traditional liberal sense, control of genomes should fall to those individuals. For many researchers, focusing on the individual and their rights impedes biomedical progress and medical benefits for all of humanity, and so should change. For example, through its MeForYou campaign launched in the spring of 2013, the University of California, San Francisco (UCSF) seeks to transform societal expectations about the use of biological material and data.[62] Directly preceding the launch of this campaign, then chancellor of UCSF and former president of product development at Genentech, Susan Desmond-Hellman, called for a "new social contract" in which individuals will not expect privacy but rather the opportunity to share their personal clinical data.[63] She argued that this shift to sharing would empower researchers to make medical breakthroughs. The MeForYou website goes further, proclaiming that "the deadliest diseases could be prevented or cured by this website." Viewers are given the chance to take part in the anticipated social, ethical, and medical transformation by following the simple instruction: "Think of the person you care about the most. Take or select a picture of yourself, and enter that person's name to dedicate yourself to their health."[64]

The simplicity of this injunction leaves many questions unanswered. In this and the many other domains where citizens today are called to reconsider their obligations to their family, community and nation, it is far from clear how to proceed. Consider the case of political theorist Michael Sandel's enormously popular Harvard undergraduate course and accompanying book, *Justice: What's the Right Thing to Do?*[65] Sandel calls his students and readers to consider their "mutual obligations."[66] However, as the sociologist Amitai Etzioni notes, Sandel does not address what one should do when one's obligations conflict.[67] The lack of a response to inevitable conflicts limits the notion of justice put forward by

Sandel, as well as the practicality of Desmond-Hellman's appeal. A call to share data on the grounds that it will help others is far from a resolved issue of justice. This is so partially because sharing information deemed personal or private raises long-contested questions of law and social order. How we decide what to make public and share, and what to exclude and for what ends, is an issue that underlies much of property and privacy law in liberal democracies and raises fundamental questions: What is a thing of value? Can it be a person's blood, DNA and tissues? If so, who can own and profit from them?

Concerns about the right to control—or own—the biological constituents of oneself connect to long-held and deeply engrained liberal commitments to self-governance. Our bodies, ourselves. This is not just the mantra of the women's health movement.[68] It is a founding principle of liberal thought in the West. John Locke, writing in a classic founding text of liberalism, *Two Treatises on Government*, proclaimed: "Every man has a Property in his own Person."[69]

However, in the wake of massive societal investment in the development of DNA initiated by the Human Genome Project, this principle faces new challenges. The US Supreme Court took up some of these challenges in its 2013 closely watched 5–4 ruling to allow DNA collection from arrestees. While some of the justices believed societal interests in DNA now outweigh a person's property right in their own person, others feared that taking DNA without a clear overriding societal benefit threatened the fundamental Fourth Amendment right to privacy. In his argument against allowing police officers to take cheek swabs in order to collect DNA, Justice Scalia invoked the Revolutionary War and framers of the US Constitution. "The proud men who wrote the charter of our liberties," he said, "would not have been so eager to open their mouths for royal inspection."[70]

The ability of DNA sequencing machines to transform DNA into data introduces additional questions about the constitution of the value of bodily tissues and their just allocation. While in the body of an individual, few would deny blood and DNA is of value to that person. However, once the tissue is transformed into data many argue just the opposite: the data have no value to individuals alone, but potentially great value when shared. In aggregate, they may help to identify new drug targets, new diagnostics, and ultimately new medical treatments that could cure cancer and other diseases. It is on the basis of this belief that institutions today ask persons to give their DNA and medical data to biomedical researchers: keeping it leads to nothing; sharing it leads to valuable things.[71] Indeed, sharing data and DNA is considered so central to biomedical progress that in some cases universities and health providers no longer ask permission. Instead, as I have learned in recent trips to see my own doctor at UCSF, sharing one's data and tissues are the Terms and Conditions of Service.[72]

UCSF's MeForYou campaign and its Terms of Service are not unique in their approach. As the chapters ahead will bring to light, the effort to reshape how individuals and collectivities understand their rights and obligations to share their tissues and data formed the hallmark of all initiatives to make the human genome meaningful in the decade following the HGP. As I chronicle in the chapters ahead, over the course of the decade, governments, health care providers, entrepreneurs, and researchers asked individuals and communities around the world to contribute their blood, tissue, and DNA less on the grounds that it would help them as individuals and more because it would benefit their community, nation, and humanity itself. These requests by researchers for access to tissue and data coincide with widespread national and international debates about how to regenerate value in a world in financial crisis where conventional natural commodities—oil, coal, timber, fish—deplete, and data deluges. Many now hope and believe that data will provide a new source of value that will be widely beneficial.[73]

Yet whether human genomics or other "big data" sectors will yield substantial benefits for human understanding and well-being is a point of contestation. When asked in July 2010 what we had learned from sequencing the human genome, Craig Venter, leader of the private effort to sequence the human genome, responded: "Very little." He went on to explain: "We couldn't even be certain from my genome what my eye color was. Isn't that sad? Everyone was looking for miracle 'yes/no' answers in the genome. 'Yes, you'll have cancer.' Or 'No, you won't have cancer.' But that's just not the way it is."[74] Writing in the same year, Elaine Mardis, codirector of the Genome Center at the University of Washington, asked in the pages of *Genome Medicine* what the value of a $1,000 genome would be if it takes $100,000 to $1 million to begin to interpret it.

If the news in the genome era was that the genome would be the secret of life and would cure cancer, the news in this postgenomic era is that we don't know how it will do this.[75] The value of knowing the sequence of the human genome turned out to be far from obvious, even to leading genome scientists.[76] Rather than reveal meaningful knowledge about life itself, genomics instead has given life to a deluge of data.[77] How to make anything of value out of this data is now, quite literally, the million dollar question.

This is not to deny the tremendous powers of the techniques and data that arose from sequencing the human genome. The many innovations arising from the HGP have led to numerous fundamental new discoveries in evolutionary biology, genealogy, conservation science, and ecology. Powerful high-throughput DNA sequencing techniques as well as the existence of the human genome sequence allowed genome scientists to sequence ancient Neanderthal DNA, and to begin to answer questions about whether gene flow occurred between Neanderthals and anatomically modern humans. Ancient DNA analysis also has allowed analysis of changes in genetic variability over time, provided insights into evolutionary relationships between species, and provided a means of testing hypotheses about the relationship between environmental events and evolutionary changes in populations.[78] Finally, the ability of genomic techniques to identify individuals and to discern relationships between individuals and groups has led to countless new practices: from new ways of doing genealogy and forensics work, to new ways of identifying which dog owners are not cleaning up after their furry companions.[79] All of these novel modes of knowing and acting face challenges—from contamination to sampling and interpretive biases—but it would be hard to deny their power to open up new areas of inquiry and to inspire new understandings of past, present, and future lives on Earth and beyond.[80]

Yet when it comes to the much-promised biomedical insights, genomics has been less than conclusive.[81] Most genome scientists promise that a biomedical revolution is still on the horizon. Yet many other scholars, as well as those asked over the course of the decade to donate their samples to genomic research, worry that the value of genomic data for improving medical care is more hype than reality, benefiting little more than the newly emergent digerati.[82]

In a moment when the divide between the rich and poor is growing and questions about the value of big data is widespread, this perception that advances in big data fields like genomics might benefit only the technological elite is not unique.[83] In the European Union in the spring of 2013, a debate over data privacy witnessed Google, Facebook, Yahoo, and other leading American information technology companies lobbying EU lawmakers to loosen privacy laws on the grounds that to not do so would be to hurt businesses already suffering from recession. However, lawmakers and citizens pushed back. They pointed to former US National Security Administration (NSA) contractor Edward Snowden's revelation of extensive NSA spying as evidence that more than just economic value was at stake, and that there might be greater value in protecting the privacy rights of their citizens and national security.[84] On the streets of San Francisco, in my own neighborhood, protesters dressed in one-piece clown suits block Google buses and challenge the idea that this mode of transportation is only doing good through providing a "green" commuting option for Silicon Valley employees.[85] The buses, they argue, are also helping to make San Francisco number one in income inequality through facilitating the move of high-tech, high-paid workers into the city's Mission District, forcing out the artists and working class who historically have lived there.[86]

Genomics is a site where these contemporary concerns, tensions, and aspirations converge.[87] What is the meaning of all this data collected on our behalf, and is it valuable in democratic societies? If so, who should be entrusted with it, and who can ensure that it does not just benefit the powerful elite? Many still hope that science and technology will succeed where other dominant social institutions have not, transforming data into new things of value that will form the basis of trusted systems of government and a more just and sustainable world. The biological and biomedical sciences, they predict, will play a particularly important role. As science and legal studies scholar Sheila Jasanoff has argued, we live in an age of bioconstitutionalism in which scientific understandings and alterations of life play central roles in constituting meaningful lives, and the rights and goods that sustain them.[88] However, efforts to render the human genome meaningful demonstrate that even the life sciences are not immune to concerns about the worth of data. These questions about the just constitution of meaning and value after the human genome, after the world financial crisis, and in the midst of deluging data and eroding trust in dominant institutions animate what I am labeling the *postgenomic condition*.

The Postgenomic Condition

In coining the *postgenomic condition*, I seek not to elevate genomics as a special case but rather to situate the questions and concerns it raises in a long history of efforts to understand the role of science and technology in augmenting or undermining conditions for life, thought, and politics.[89] It is a riff on Hannah Arendt's *The Human Condition* and Jean-François Lyotard's *The Postmodern Condition*.[90] Arendt and Lyotard are two of the twentieth century's most articulate and reflective commentators on the ethical, political, and philosophical questions raised by post–World War II investments in science and technology.[91] I invoke them not out of nostalgia, or to recall a more enlightened past, but rather to clarify the broader issues at stake as the products of the life sciences and biotechnology—such as the human genome—today attempt to become the grounds for democratic and just action.

Both Lyotard and Arendt wrote in the wake of developments crucial to the emergence of human genomics as a field of study: Arendt, in 1958, five years after the report of the molecular structure of DNA; Lyotard in 1979, two years after the sale of the first personal computers. They both raised concerns about the effects the rise of informatics and mathematics—which accompanied the rise of molecular biology and computers—would have on knowledge and politics. Arendt argued: "The trouble concerns the fact that the 'truths' of the modern scientific world view, though they can be demonstrated in mathematical formulas and proved technologically, will no longer lend themselves to normal expression in speech and thought. . . . We do not yet know whether this situation is final. But it could be that we . . . will forever be unable to understand, that is, to think and speak about the things which nevertheless we are able to do."[92] Today, as an increasing number of genome scientists contend that the algorithm should replace the scientific hypothesis as the starting point of scientific inquiry, math and formulas are again on the rise.[93] Half a century ago, such developments prompted Arendt to ask fundamental questions about the conditions of knowledge and politics. She worried that human beings would lose their ability to engage in meaningful speech, that which she believed distinguished "acting men" from "performing robots." Speech, Arendt believed, gives us the stuff of stories through which we constitute a common world.[94] If meaningful speech went, so too, she feared, would politics and a public realm of debate, both of which she deemed essential to humanity. Ultimately, Arendt questioned whether the human condition is compatible with modern forms of science and technology. Writing in the wake of the Holocaust and the murderous entanglement of the natural sciences and society, she speculated that the answer might be "no." In any case, she believed that post–World War II investments in the natural and physical sciences should prompt questions about how we could retain our ability "to think what we are doing"—which for Arendt was the starting point of political life.[95]

Lyotard writes twenty years later. By this time, Arendt's vision of a society that "needs" machines to make sense and administrate has come to pass. The "computer age" has arrived. For Lyotard, as for Arendt, this rise of "thinking" machines raised questions about the nature of knowledge and politics. Since the dawn of the age of reason in the eighteenth century, Lyotard notes, modern political life took rational knowledge as its guide on the grounds that such knowledge either represents universal *truth* (that which transcends everyday earthly existence and connects one to God) or *justice* (that which provides freedom here on earth). In the postmodern condition, he posited, the grounds shift. Knowledge is no longer the product of trained minds but the output of computers.[96] Computers enable automation. Automation undermines reflective thought. It instrumentalizes and commodifies, allowing "thinking" to become a product of machines that can be bought and sold, and science "to be completely subordinated to the prevailing powers."[97] The result, he concluded, is that knowledge is unhinged from both its claim to truth and to justice. Easily harnessed for specific financial and political ends, computer-mediated thought ended any illusion that economic and political interests had no stakes in questions about what we should know and how we should know it. It became clear that knowing as an activity fosters the well-being of some at the expense of others.

I write nearly three decades later. Today the computerization of society meets its biologization in the form of genomics. One hallmark of genomics is that it thoroughly integrated computers and informatics into the biological sciences.[98] For the sequencing of the human genome to be feasible, machines needed to replace the human beings, such as myself, who once poured the sequencing gels and recorded and interpreted the resulting sequence data. Three billion bits of data were simply too many for the unaided human mind to comprehend and process.[99] Machines could and should replace it. They could, Venter argued, work more efficiently and "eliminat[e] the need for thousands of workers."[100] At the end of the day, both the public and the private efforts to sequence the human genome sought to purchase as many 3,730 ABI (Applied Biosystems Inc.) sequencing machines as possible in order that they might more quickly sequence the human genome.[101]

After the HGP, many genome scientists hoped this industrial technological arms race would end and scientists would return to the task of focusing on creating meaningful insights about life. However, it soon became clear that advances in sequencing technology would continue to generate the most money and excitement. The premier sequencing project—the Human Genome Project—might be complete, but many believed medical breakthroughs required radically decreasing the cost and increasing the speed of DNA sequencing. Only these innovations, they argued, would allow millions and eventually billions of human beings to have their genomes sequenced. Thus, in 2004 the NIH began the Advanced Sequencing Technology Awards grant scheme, which over the next decade provided $230 million worth of seed money to spur innovation in sequencing. Venture capital (VC) funding and an X prize soon followed. DNA sequencing became a multi-billion-dollar business. Advances in "Genome Biology and Technology" meetings transformed into lavish affairs held on island resorts where guests enjoyed heated pools and Jacuzzis, open bars, and lavish spreads of food all underwritten by ABI and its competitors.[102]

DNA sequencing companies soon became the darling of the tech world, exploding Intel co-founder Gordon Moore's law that that computing power doubled every two years and the price therefore halved. Sequencing powers increased five times as fast, doubling every three months.[103] Despite the initial post-HGP efforts to return focus to fundamental questions of biology and medicine, this startling demonstration of technological prowess captured the limelight and VC dollars and created a powerful force that few could escape. As we will see, even initiatives that began with clear intentions to focus their energies on the interpretation of genomic data and the creation of medical diagnostics and treatments eventually returned to the goal of improving sequencing machines.

Genome scientists understood the threat that the large sums of money at stake in sequencing posed to the communal and democratic ethos propounded by the US president and embraced by leaders of the HGP. They resolved to fight the forces of capitalist accumulation and to help ensure that the resources that flowed to sequencing did so in a more equitable fashion. Thus, leaders at NIH distributed grant funding to companies who could compete with ABI. Their goal: break the ABI monopoly.

ABI did falter. The DNA sequencing monopoly did not. Today, one powerful company again dominates: not Applied Biosystems, but Illumina. In 2015, Illumina was valued at $28 billion, and produced more that 90 percent of all DNA data.[104] Perhaps more importantly, the company seeks not just to sequence DNA but to own the means of interpreting it. A look at the company's home page provides a glimpse into its aspirations to dominate all aspects of postgenomics, from forensics to the microbiome to oncology to reproductive health.[105] While many hoped that in the era after the Human Genome Project the field would diversify not only the genomes sequenced but also genome sequencers, it is here that we have arrived: a genomics in which one powerful corporation and its machines lie at the center.

The human genome promised to reveal fundamental new knowledge about life and what it means to be human. Whether or not it succeeded is a matter of much academic writing and debate.[106] What is less contested is that the initiative moved sequencing machines, automation, and the cultures and economies of innovation into the heart of the life and biomedical sciences. As the stories that follow illustrate, it removed human beings from the center of buildings in the life sciences, replacing them with computers and other automated machines. This is the irony of term "human genome project." It is also one source of the fear that the Human Genome Project threatened the human condition in the Arendtian sense.

Concerns about the potential inhumanity of genomics also would be raised by comments made by James Watson, the co-discoverer of the double helical structure of DNA, and the first director of the Human Genome Project, in an interview with the London paper the *Sunday Times*: "[I am] inherently gloomy about the prospect of Africa [because] all our social policies are based on the fact that their intelligence is the same as ours—whereas all the testing says not really."[107] Claims such as these by prominent leaders in genomics and in the academy, as well as the vast influx of venture capital into the life and biomedical sciences that accompanied high-paced innovation in sequencing, undermine claims that "the human genome" represents a universal good.

One might predict that the drive toward automation and the concentration of resources, combined with statements widely viewed as racist made by a prominent leader, would undermine the fields' claims to truth and justice. Yet unlike what Lyotard predicted, in genomics we have witnessed a rise in claims to justice. As we will see, in the decade after the Human Genome Project, genome scientists responded to fears that genomics would lead to new forms of racism and biocolonialism by arguing that their science was liberatory and advanced causes of access and equality.[108] They celebrated genomics' power to undo social concepts of race.[109] They championed calls for open and public data.[110] They sought new rights for citizens.[111] In short, in *the postgenomic condition*, far from declining, grand narratives about the link between science and a more just world became central.

Threatened on the one side by an informatic purgatory, and on the other by charges of racism, justice and democracy filled the breach.[112] How did this happen? How within genomics did efforts to democratize and to create justice become harnessed to the rise of the machines? In the chapters that follow, I explore what this remarkable turn of events reveals about the conditions for creating knowledge, ethics, and justice in this time after the human genome.

The Problem of the Post: What Comes after Value?

One central aspect of these conditions is the question of the "post" that is implicit in the phrase, the postgenomic condition: What comes after? In a world that increasingly revolves around technological innovation, many operate in the present with an eye to the future, always seeking the latest greatest technical platforms. Even before the iPad mini was officially announced, I found myself wondering what would come after? Is 7.9 inches the right size? What about 6? Do I invest?

Genomics exemplifies these trends. In 2007, personal genomics companies genotyped genomes with an Affymetirix SNP chips and created private websites that could be accessed by customers. Today, they outsource whole genome sequencing to the lowest bidder and use the Amazon Cloud to enable access. Getting caught with the wrong set of machines, on the wrong platform, is a constant concern. An endless process of production and consumption takes hold.[113] Nothing endures; all is "in-formation."

As Arendt argued in *The Human Condition*, without endurance "no thing" sticks around long enough for us to gather around for democratic deliberation or knowledge-making.[114] Informed decisions become outmoded as there is no *thing* around which we can gather and deliberate. Things, after all, require our attention and concern. Recall, as the French theorist of science and technology Bruno Latour reminds us, the word *thing* derives from the Old English word "ding," which means a meeting or assembly.[115] A thing, or a gathering, arises around a matter of concern.

While Latour coined the phrase *matters of concern* to call our attention to this way in which our concerns shape what things exist in the world, Maria Puig de la Bellacasa created the phrase *matters of care* to help us see the active work that must be done to constitute things, and the political and ethical questions at stake in the distribution of this labor.[116] The creation of things, she argues, raises a fundamental question: Whose cares and concerns should matter and give rise to the things in the world around which we gather? The debate over whether the genome is meaningful—something or "no thing"—arises from this critical matter: Whose concerns, and thus what kinds of things, will make up the world in the opening decades of the twenty-first century? Will a genome be one of those things?

This is a very different kind of question from the one raised during the era of the Human Genome Project. When most believed the sequencing machines were generating something of great value—the human genome—it made sense to talk about who controlled it, who could own it, and how its power might be used to include and exclude. However, in the postgenomic condition these machines now often produce sequence of unclear value. How do we decide when the genome sequences of passenger pigeons, speckled mousebirds, and Parnell's mustached bats are of high enough quality to be of use to the scientific community?[117] Investments in the HGP and biomedicine secured enough researchers interested in the human to ensure that the human genome would be sequenced enough to eliminate most errors. Now that sequencing has extended to thousands of species, for whom there may only be a handful of interested researchers, how to decide when sequence data is good enough to be of value often remains unresolved. Further, the meaning of the sequence is frequently far from clear. This is particularly the case in medical genomics, where there are still few genetic variants with clear clinical meaning.[118]

How can we think and judge when so much is undecided? As a social scientist involved in genomics research in parts of the African continent explained to me, when people know what the value of something is, they can weigh up and decide whether they want to take part in it. For many of the people with whom she works the value of genomics is hard to discern. How then can they decide whether to participate in genomics?

Ethics, Justice, Knowledge after the Genome: From After-Thought to Stories

These questions lie at the heart of *the postgenomic condition*. To address them, in the chapters that follow I tell a series of stories around which we might gather to understand and discuss efforts to make the human genome meaningful for a wide and diverse range of lives. Stories lie between us; they are quite literally of "inter-est."[119] We share stories, as Arendt argued, to build a common world from capacities to see it through the eyes of others. In today's world, as in Arendt's, stories are the necessary companions of the formulaic languages of computers. While codes and algorithms can compel action, they do not help us to critically think and understand why or how we should act. They are instrumental, but not revelatory. They promise control, not the freedom to contest. They process data, but do little to cultivate meaning. For this we require stories, stories that help us "to think what we are doing," to critically engage with and respond to that which lies between us.[120]

There are no shortages of articles and books that tell us of the great ends and great troubles that might result from human genomics, but there are scant few that help us to think about what we are doing along the way, and how to respond. This book fills this gap with stories not about genomic heroes and villains, utopias and dystopias, but about the wide webs of humans and machines, hopes and disappointments, histories and futures that make up the postgenomic condition. In each, the growing mandate to collect genomes and to provide open access to genomic data in the name of medical breakthroughs meets the particular bodies, histories, and constraints of the people asked to provide their genomes. The aspirations of openness and unbounded freedom meet the lived realities of constraint, choices, and exclusions. Informatic openness bumps up against biomedical and liberal democratic commitments to privacy, creating fundamental questions about what constitutes just and ethical governance. Stories of these encounters take us out of all-too-familiar celebrations or condemnations of genomics to engage the ambiguities, paradoxes, and dilemmas posed by the on-the-ground realities.

While the chapters are ordered chronologically so as to follow the unfolding of postgenomics over the course of the decade, at the heart of each is a core concept in liberal democracy—information, inclusion, the people, persons, property, privacy, and public. My contention is that different liberal concepts were salient at different moments. I begin with information, as this concept structured the development of genomics from the start. Inclusion emerged later, and persons is a recent newcomer. The rise of each brought new hopes for forging a meaningful genomics that could live up to the promise of understanding and supporting the human species and all its many companion species.[121]

My own travels for the book followed these hopes. The daughter of a former Jesuit priest who espoused the ideas of Pierre Teilhard de Chardin—a fellow Jesuit who believed that humans were evolving toward a maximum level of consciousness—I learned early of genetics and its promise to advance the human condition.[122] Yet despite my father's urgings, I had more immediate interests: ecology, philosophy, and politics. However, at the end of my undergraduate training, I circled back to genetics. The Human Genome Project had just launched, and I felt the excitement that surrounded the powers of molecular techniques to reveal fundamental new truths about life broadly understood. Caught up by that excitement, I traded my field study waders for pipettes and began to sequence genomes. I learned DNA sequencing when it was still humans, not machines, who did most of the work. Inspired by the promise to not only understand fundamental new biological truths, but to further human rights work, I was accepted for graduate study in the Molecular, Cellular, and Developmental Biology Department at the University of California, Berkeley. I hoped to work with Mary-Claire King, who at the time was pioneering genetic techniques for identifying the children who had been taken from their families during the Dirty Wars in Argentina. King also just had coauthored a call for the Human Genome Diversity Project (HGDP).[123] I found both compelling, promising as they did to bring together biology, history, politics, and social justice.

Yet, ultimately, I decided my questions spilled outside the laboratory walls, and that I wanted, *à la* Arendt, to attempt to think and write what we were doing as so many turned their energies toward sequencing DNA. Instead of staying in California, I headed East, obtaining my doctorate at Cornell University in a field that just then was emerging: science and technology studies.[124] There I explored my interest in human genome diversity research as a meeting place of biology, politics, and history, and would go on to write a book about it.[125] I also kept my strong ties to and interests in the practices of molecular biology and genomics, links that would be essential to my ability to understand and chronicle the emergence of human genomics.

What dramatically changed was my understanding of the nature and meaning of human genomics. While for some the Diversity Project promised fundamental new truths about human history and evolution while fighting racism, for others it represented a threat to their histories and livelihoods.[126] I was startled and puzzled by the clash. How could this project that I had gravitated toward for its potential to address broad human problems and advance the cause of justice be viewed by some as the next wave of scientific racism? While it offered to build meaningful lives for some (for example, my younger self), for many whose genomes Diversity Project scientists sought to collect it threatened to reduce their existence to what DNA could reveal about their past, threatening their claim on a future.[127] This tension both troubled and interested me.

While unsettling, this tension also revealed to me the power of human genomics to generate insights about life in its broadest sense. The field not only addressed fundamental questions about evolutionary history and disease, it also brought into focus the nature of contemporary societies and politics. Human genomics, I discovered, meant more than I ever imagined. It meant a struggle over how we might understand human beings not just for treating them for diseases but for settling disputes over property, identities, and resources. It meant the rise and fall of careers. It meant decisions about what futures to invest in and which pasts to redress. It meant choices about which lives to know and care for, and which to ignore and to let perish. Studying genetics and genomics, as my father had suggested, could indeed expand consciousness.

Yet I still hoped it might also mean some things over others: a world that includes more than it excludes, social justice and not racism, openings rather than enclosures. I chose each of the field sites in this book because all of them are enmeshed in similar aspirations. From Tuskegee, Alabama, to Edinburgh, Scotland, from Mexico City to Silicon Valley, over the course of the last decade I met genome scientists, social scientists, informaticians, genetic counselors, bioethicists, social activists, lawyers, entrepreneurs, and policy makers who shared a desire to ensure that human genomics made good on its original promise: that it would represent all humans; that it would help build a better future for everyone. In my interviews with them, I learned about the specific challenges they faced, the unexpected lessons learned, and the unchartered territories that lay ahead.[128] While their commitments to the promise of human genomics so elegantly articulated by President Clinton were clear, how to uphold that commitment on the ground in specific projects often remained far from evident. Yet much would be learned. While their original goals were often illusive, always they made pivotal contributions to addressing questions about how we can know and govern in times when technoscience garners disproportionate resources, the gap between the rich and poor grows, and liberal democratic modes of thought and action often provide inadequate responses to worlds in need. Together, their stories invite—and I hope will help to create—a much broader public discussion about how, after the human genome, we might forge ethics, justice, and knowledge on a scale that can be lived and shared, not merely hoped and dreamed.[129]

To appreciate these stories, it is important to first understand the hopes and dreams from which the idea to sequence the human genome first arose. How did liberal dreams of good government grounded in freely accessible information become harnessed to genomes, and how did the resulting notion of genomic "information" produce the postgenomic problem of meaning? It is to this question that I now turn.

2 THE INFORMATION OF LIFE OR THE LIFE OF INFORMATION?

Although the sequence is in one sense the ultimate map, it is also much more than that: it is also the biological information itself. When we finish collecting the sequence we will have the hieroglyph of biology in our hands, even if we don't at first understand it all.
—John Sulston and Georgina Ferry, *Common Thread*

On August 6, 1945, the US government dropped a uranium bomb on the city of Hiroshima. Three days later, a plutonium bomb exploded over Nagasaki. Tens of thousands lost their lives instantly. Thousands more died more slowly from burns and radiation sickness. Countless others perished in the decades to come from the ongoing effects of the radiation. Yet despite this massive loss of human life, many celebrated the bomb.[1] Rightly or wrongly, they argued it ended World War II. While it killed hundreds of thousands, they believed it stopped the bloodshed. Riding on this perceived success, US government and industry investments in science and technology grew massively directly following the war's end. However, beginning in the 1960s, as deaths in Vietnam mounted, public sentiment turned.[2] Controversies broke out over the toxic and deadly effects of Agent Orange, radiation exposure during the atmospheric nuclear tests of the 1950s, and exposure to mutagenic chemicals. Life and death, many began to suspect, turned not on a clear moment of physical impact—the explosion, for example, of an atomic bomb—but on forces more pervasive and less visible.[3] While a bullet or a conventional bomb created clear casualties, the atomic bomb maimed and killed its victims in unknown new ways: slowly, invisibly, over time.[4]

The Human Genome Project was born out of the problems of knowledge and justice produced by these ghosts of the atomic age. Worried that the controversies over the effects of atomic radiation would lead to lawsuits, the US Congress set out to create "the evidence of things not seen."[5] They wanted to know what could and could not be known about the mutagenic effects of radiation. In December 1984 in a ski lodge in Alta, Utah, the Department of Energy and the International Commission for Protection against Environmental Mutagens and Carcinogens convened experts on DNA analytical methods to determine the technical capacities required to directly detect mutations. Their conclusion: a large, complex, and expensive program to improve DNA analysis was needed.[6]

While the HGP began with this concern to redress unjust injury and loss of life resulting from the harnessing of the physical sciences to the ends of warfare, a pursuit of justice older and deeper shaped its evolution: the quest to create access to the information needed for people, and not tyrants, to rule. Like the "father of the bomb," theoretical physicist J. Robert Oppenheimer, the founders of the Human Genome Project believed in the Enlightenment notion that information and knowledge are the cornerstones of more rational, and thus more just societies.[7] DNA, they believed, harbored that information in the biological world. Its sequence, HGP leader John Sulston proclaimed, was "the biological information itself."[8]

Yet like the physicists who sought to discover the first principles of the physical universe, genome scientists who studied what they believed to be the building blocks of the biological world quickly found themselves embroiled in larger political economic forces that troubled their long-held and cherished beliefs and practices. The so-called secret of life turned out to also promise a pot of gold.[9] Thus, according to the dominant popular accounts, genome scientists quickly found themselves on the front lines of a battle to defend "the most fundamental information about humanity" from a corporate takeover.[10] Their moral economy—guided by norms of openness and communalism—struggled to survive in the face of the growing power and widening influence of a capitalist economy.[11] In these stories, moral judgments are clear: John Sulston, leader of the public sequencing effort, is a hero; Craig Venter, the former NIH scientist who led the private effort, is a villain.

Based on interviews with the main architects of the Human Genome Project, as well as their autobiographical accounts, this chapter tells a different story.[12] In this story, the forces at work are not so clearly visible. No distinct lines between good and evil, public and private, the moral and immoral can be drawn. The relationship between a novel form of technoscience—this time DNA sequencing—and life and death is once again uncertain. Justice does not hinge on the actions of individuals—whether heroic or abhorrent—and blame cannot easily be ascribed. Instead, all the actors—whether they worked for a company, a venture capital firm, or a public government—find themselves enmeshed in a fundamental struggle over the values and meaning of science and information in the wake of post–World War II investments in communication technologies, post-Vietnam questioning of the value of technoscience, and the post–Cold War rise of informatic capitalism.[13] What is the value and meaning of an endeavor that requires an ever-growing number of automated sequencers to displace humans and that consumes large amounts of reagents and capital? Who and what benefits from these informatic and automatic infrastructures? Despite the valiant efforts of leaders of the HGP to defend and advance public-funded science, many on the ground feared that genomics ushered in a technocratic and capitalist mode of producing information, one in which computer-run machines designed to increase speed and efficiency replaced humans who sought knowledge and justice.

Their concerns did not go un-storied. Many genome scientists wrote accounts of the Human Genome Project that brought to the fore these deeper structural transformations that unsettled the simple story that access to genomic information was good ethics and good science.[14] By bringing these accounts in conversation with the dominant account of the Human Genome Project, in this chapter I forge the vistas required to understand these deeper complexities from which the postgenomic problem of meaning arose. I bring into focus one central question this problem posed in the decade after the HGP: How can we know and act ethically in a world where life becomes information, information becomes capital, and capital is equated with freedom?

From Bermuda to Bethesda: Forging Genomics' Founding Stories

On February 12, 2001, leaders of the Human Genome Project and their supporters packed the Masur Auditorium on the NIH campus in Bethesda, Maryland, to celebrate the simultaneous publication in *Science* and *Nature* of the draft human genome. Fittingly, it was also the birthday of Abraham Lincoln. As the rest of the United States celebrated their sixteenth president—remembered for saving the Union and assuring that "government of the people, by the people, for the people, shall not perish from the earth"—members of the seven-country, twenty-institution public consortium to sequence the human genome celebrated saving the human genome for the people.[15] The "bad guys," announced Jim Watson, co-discoverer of the double helical structure of DNA and first director of the Human Genome Project, did not win. Despite the effort of Craig Venter and his corporate sponsors, the genome would remain, in the words of the leader of the United Kingdom's human genome sequencing effort, John Sulston, "the common heritage" of all people.[16]

At the celebration in Bethesda, Francis Collins, born in the Shenandoah Valley to a father who collected folk songs, sang a song performed to the tune of Woody Guthrie's "This Land Is Your Land":

> This draft is your draft, this draft is my draft,
> And it's a free draft, no charge to see draft.
> It's our instruction book, so come on have a look,
> This draft was made for you and me.[17]

Collins followed in a long line of folk singers—from Bob Dylan to Bruce Springsteen—who sought to revive Guthrie's anticapitalist political message. As Guthrie wrote in one verse of his much-beloved song:

> There was a big high wall there that tried to stop me;
> Sign was painted, it said private property;
> But on the back side it didn't say nothing;
> This land was made for you and me.[18]

The song, Collins explained, "[summed] up why we did all this and what some of our hopes are."[19] The true significance and meaning of the Human Genome Project, he argued, was its defense of public science and the freedom of information.

All in the audience knew the story. In the mid-1990s Craig Venter and his corporate backers as well as Myriad Genetics attempted to patent the human genome sequence: first expressed sequence tags (ESTs) and then entire genes (for example, *BRCA1*). In Bermuda, in February 1996, scientists leading the public effort to sequence the human genome from the United Kingdom and the United States agreed to fight off these enclosures and forged a much-celebrated historic data-sharing agreement. At that meeting, Sulston famously scribbled on a white board:

Automatic release of sequence assemblies > 1 kb (preferably daily)
Immediate submission of finished annotated sequences
Aim to have all sequence freely available and in the public domain for both research and development, in order to maximize its benefits to society.

These statements, with some slight modifications, became known as the Bermuda principles. Sulston credited them with playing the pivotal role in saving the human genome from patenting and ensuring its preservation as a public good.[20] Ratifying them, Collins argued, "was one of the defining moments of the HGP."[21] They served as an important rallying call for the public effort when, in the colorful words of *Drosophila* geneticist Michael Ashburner, the scélérates—the dreaded "men in suits"—landed in 1998, giving Venter the venture capital needed to form Celera. With $300 million in funding and a massive army of sequencing machines, this company, named for the Latin word for speed, threatened to sequence the human genome before the public effort. However, the bearded, folk-song-loving genome scientists rounded up their troops and worked ceaselessly to save the human genome from private enclosure.[22] Jim Kent of the University of California, Santa Cruz, reportedly spent day and night in his garage writing code, stopping only to ice his wrists.[23]

At stake, many genome scientists argued, was not just the human genome, but the future of science itself. Wrote geneticist R. Scott Hawley in his epilogue to Ashburner's account of the sequencing of the fly genome, *Won for All*: "Science can only progress if knowledge, and the material resources required for acquiring knowledge, are freely distributable to our colleagues—be they collaborators or competitors."[24] The leaders of the HGP publicly celebrated this ethical commitment to the free flow of information and knowledge.

Human Genomics: A Moral Economy?

In placing the free flow of information and knowledge at the moral heart of genomics, Hawley, Sulston, and Collins drew upon an understanding of science that originated a half century earlier in the work of the founder of the sociology of science, Robert Merton. Writing in the shadows of the rise of fascist governments in Europe, Merton argued in his "Notes on Science and Democracy": "To the extent that a society is democratic, it provides scope for the exercise of the universalistic criteria of science." Central among those criteria was the norm of "communism." Science, Merton explained, was a communal good and its results should reside in "the public domain." "Secrecy," he wrote, "is the antithesis of this norm; full and open communication its enactment."[25] Merton contrasted the open communication of science with the private enclosures of technological inventions. The US Supreme Court, he observed, ruled in *U.S. v. American Bell Telephone Co.*, in 1897: "The inventor is one who has discovered something of value. It is his absolute property. He may withhold knowledge of it from the public."[26] This conception of technology as "private property," Merton argued, posed a threat to science.

For decades, Merton's understanding of science as a beacon of openness held sway among natural and social scientists. Directly after World War II, Vannevar Bush, then head of the Office of Scientific Research and Development, famously argued for the importance of opening up and publishing previously classified military information.[27] During the 1950s and 1960s, scholars mobilized the ideal of open science in their fight against governmental secrecy and Big Science.[28] When in the mid-1990s genome scientists invoked the same ideal to organize and support the Human Genome Project, they drew upon this long and venerable history of the idea that among scientists information should flow freely.[29]

Yet how and why public access to genomic information came to occupy a central place in genomics is far from a historical inevitability, and in no way a simple tale. To begin with, genomics does not fit easily into Merton's ideas about the distinction between science and technology. From its inception, technological innovation was integral to the field. Any attempt to draw a line between science and technology within genomics, as the sociologist of science Stephen Hilgartner has noted, is "empirically elusive."[30] This ambiguity troubles any easy invocation of Mertonian norms. While science might be governed by an ethos of openness, from its very conception the *techno*science of genomics raised concerns.[31] In 1980, nearly one hundred years on from *U.S. v. American Bell Telephone Co.*, the Supreme Court's *Diamond v. Chakrabarty* decision to uphold a patent on a genetically engineered bacterium indicated that the notion of "technology as property" would persist into the era of biotechnology.[32]

This meeting of the private property regimes of technological innovation with the Mertonian norm of scientific openness created a formative tension that powerfully shaped genomics from its start. From very early on, leaders of the HGP recognized that if they were to sequence the human genome within their lifetimes they needed proprietary automated machines. Like Celera, they sought large numbers of the commercially produced Applied Biosystems Inc. (ABI) automated sequencing machines.

For John Sulston, the UK leader of the HGP known for his efforts to keep the human genome sequence public, this created a formidable challenge. In the pages of his autobiography, *The Common Thread*, he devotes several pages to describing the dilemmas he faced as he sought to take advantage of the powers of speed and efficiency the ABI machines offered while not becoming entrapped in the company's property regimes. He came of age as a scientist, he explains, in the celebrated Laboratory of Molecular Biology (LMB) then led by one of the founders of molecular biology, Sydney Brenner, and the co-discoverer of the structure of DNA, Francis Crick.[33] More than any of the biological breakthroughs, Sulston celebrates the social organization of the laboratory: tightly packed lab benches that encouraged interaction and sharing of ideas.[34] The productivity and success at the LMB, he explained, hinged not on computers and a capitalist economy, but on humans and a sharing economy.

As the historian of science Robert Kohler demonstrates in his history of *Drosophila* genetics, these routines of sharing have deep historical roots in gene mapping communities.[35] Thomas Hunt Morgan and his "fly boys" routinely shared stocks, a practice that scientists and historians alike widely credit with increasing the speed of their knowledge production and the broad acceptance of their work and practices. It later became standard practice across molecular biology, adopted by maize, phage, bacterial, and worm geneticists.[36] Sharing became so central to these communities, Kohler argued, that it defined insiders from outsiders. It was, he argued, "a badge of citizenship."[37]

Decades later, when scientists sought to map the human genome, Sulston argued that sharing remained fundamental to their success. As he explained to Duke University undergraduate Lina Lu in 2007: "The whole process of mapping the genome only works if people share their information. If people don't share, then it doesn't work. . . . And this actually becomes very serious with something like the human genome where the dataset is so vast and so little understood."[38] Sulston here and elsewhere makes clear that sharing is not just important politically and ethically, it is a practical necessity: "We had to work together, because nobody at that stage could do the whole thing by themselves," he recalled.[39]

Yet sharing, Sulston is quick to point out, did not rule out rights of ownership. To prevent duplication and to ensure the whole genome would be sequenced, those engaged in sequencing the human genome made agreements about who controlled sequencing particular segments of the human genome, agreements that mattered most when those segments were thought to contain important genes (such as those associated with breast cancer, *BRCA1* and *BRCA2*).[40] These de facto ownership agreements, he explained, fostered the systems of responsibility and trust needed to sequence the whole human genome. The new problem Sulston faced with the arrival of Venter and the ABI machines was *not* that they threatened to assert property rights over the human genome, but that they did so in a manner that created *private*, not *communal*, property, and that placed money, not recognition and academic prestige, at the center of the exchange.[41] In the grand narrative of *The Common Thread*, Craig Venter is an enemy because his decision to join the *capitalist* economy threatened what Kohler described as the *moral* economy of gene mappers.[42]

Yet what kind of problem is this? While a moral economy at first brush might sound better than a capitalist economy, a deeper look at the meaning of these terms reveals a more complicated story. Kohler draws his use of the term *moral economy* from the social historian E. P. Thompson, who in his 1963 book *The Making of the English Working Class*, used *moral economy* to describe the practices and norms that adjudicated distribution of bread during the food riots in eighteenth-century England. While many have interpreted Thompson as arguing that the "moral economy" of the poor during the eighteenth century was better than the laissez-faire "political economy" of the market during the nineteenth century (or, just simply, "market economy"), Thompson later clarified that this was not his intent.

> Maybe the trouble lies with the word "moral." "Moral" is a signal which brings on a rush of polemical blood to the academic head. Nothing has made my critics angrier than the notion that a food rioter might have been more "moral" than a disciple of Adam Smith. But that was not my meaning. . . . I could perhaps have called this

> "a sociological economy," and an economy in its original meaning (*oeconomy*) as the due organization of a household, in which each part is related to the whole and each member acknowledges his various duties and obligations.[43]

Yet as others imported the term into other domains, including the history of science, they opened the door to acts of judgment.[44] To describe some things as "moral" implies that there are other things that are amoral or immoral. The term carries along with it an implicit normative judgment. Praise and admiration comes to those credited with creating and supporting moral economies. And indeed, as Kohler reports, Morgan and his fly men were held in high esteem.

So were John Sulston and Francis Collins. The Queen of the United Kingdom knighted Sulston in 2001. President Barack Obama appointed Francis Collins director of the National Institutes of Health in 2009. In the dominant stories about the Human Genome Project, Collins and Sulston are virtuous men recognized as national heroes. Readers of *The Common Thread*, for example, will not miss the judgment implicit in the main narrative of the book. Venter and his venture capitalists are the villains. They threaten the right order of public science. Sulston is the hero. He valiantly fights off the villains. When Sulston spends a Sunday afternoon decrypting the ABI machines' files so that he and his colleagues do not have to depend on the company's proprietary software, his readers cheer him on.[45]

Yet these stories of the "moral economy" of the HGP engender the trouble that Thompson describes. They encourage a polemical understanding of what in reality was a much more complex story. As a principle architect of the Human Genome Project later explained to me, the story that the Human Genome Project sought to liberate the genome for all was an appealing and powerful one, but it ignored on-the-ground realties.[46]

In reality, Bermuda's policies of openness favored the inclusion of some over the exclusion of others. As one of the early innovators of genome sequencing techniques reflected in 2013: "I think that there was a not very well recognized self-interest on the part of the then dominant DNA sequencing centers. It made it materially more difficult for minor contributors to the human Genome Project to meet the Bermuda rules. The Sanger Institute had a one-hundred-person IT (information technology) staff, and it wasn't really a problem for them to funnel their data right into GenBank. It was a serious problem for me, and I resented being told that I should divert other modest resources from what I thought was the best management of our endeavor."[47] Uploading large data sets and putting them in the open domain involves lots of time and resources. It is not as easy as the push of a button. The data must be properly formatted and shepherded through the upload process. Openness required resources, and resources were limited.[48]

Does Openness Support Knowledge?

Not only did the Bermuda principle of openness pose questions about who had the resources needed to take part in genomics, it also raised questions about how to achieve the quality of data needed to create genomic knowledge. Not all agreed that openness fostered knowledge. Indeed, to the contrary one leader of the HGP claimed just the opposite:

> The core of the Bermuda model was that data should be released without any analysis whatsoever, not even quality control. . . . I think that idea is actually silly. It's not even worth building some complicated counter argument; it's just silly. It's a core principal of science that scientists need to be held to some account for the reliability of the information that they inject into the commons. You cannot simultaneously ask them to release their raw data in twenty-four hours and be accountable as far as validity. Scientists aren't ever going to function that way [as Bermuda asked them to], and they actually shouldn't.[49]

Even at this early stage of human genomics, scientists worried about the epistemic value of the data they were producing.[50]

This would not be the first time in the history of knowledge production that a commitment to open access threatened to impede the creation of valuable knowledge. As historian Adrian Johns describes in *Piracy: The Intellectual Property Wars from Gutenberg to Gates*, in nineteenth-century Britain, universal deposit laws raised fears that libraries would become "infinitely large reservoirs of triviality."[51] Cultivation and dissemination of knowledge historically and today requires extensive time, money, and attention. Thus the sense in which it should be free is far from clear and has long been a matter of debate.

While felt on the ground at the time, these deeper complexities of open access did not figure in the popular accounts of Bermuda. Instead, a popular fairy tale account developed. As one HGP scientist put it: "I think there was sort of a white knight sort of complex. It sounded good to people who didn't think about it very much, and people like to be presented as heroes."[52] Bermuda is an island. A land of dreams and ideals where one can walk around Lover's Lake and surf in the crystal blue waters and swim with the dolphins. While Bermuda continues to live on as a key founding myth of an open, responsible and enlightened genomics, life off the island revealed that the story was never that simple.[53]

Off the Island and Into the Prison

"I just heard the prison door close behind us."

Thus begins *The Common Thread*. It is Sulston's description of the moment he realizes that he has signed on to the Human Genome Project and there is no going back. Having just met with Jim Watson at Cold Spring Harbor, he is standing on the train platform at Syosset on the Long Island Railroad under a sun "harsh and bright," the sound of a prison door "reverberat[ing]" in his ear. Given the book's emphasis on the importance of openness, and its triumph over the nefarious forces of private enclosure, this is a surprising choice for an opening line. Yet the theme of the prison door clanging shut reverberates throughout the book, alerting the reader that this will be no simple tale.[54] While on one level a celebration of Sulston's understanding of the scientific life, and the power it affords individuals to think freely and to overcome powerful political and economic forces, on another it is a poignant account of how Sulston and his colleagues lost control, and along with it their long-valued communal practices of constituting trust and creating knowledge.

In the pages that follow, Sulston chronicles the transformation of his everyday work life as it moves from the small intimate spaces of intense human interaction at the Laboratory of Molecular Biology in Cambridge to the vast cavernous spaces of the sequencing machines at the Sanger Centre built in Hinxton, miles away from molecular biologists' much beloved pub, The Eagle. We learn of a life no longer marked by late night encounters in the lab, coffee time, drunken punting expeditions and Guy Fawkes celebrations, but a life lived under the pressure of keeping an army of sequencing machines producing data on schedule.[55] The needs of the machines, and not those of human beings, take center stage. They are needs that bring Sulston reluctantly, but seemingly inexorably, into a capitalist world of production, for first and foremost, genome sequencing requires massive amounts of resources: to buy machines, hire sequencing teams, and pay for a constant supply of reagents.[56] Throughout his account of the nearly fifteen years he devoted to sequencing the human genome, the amounts of money required continually threaten to overreach the capacities of public governments and private foundations, making the project of sequencing the human genome perennially vulnerable to a venture capital takeover. Sulston receives multiple offers to join private industry, offers he reports he always took seriously, even when they came from his archrival, Craig Venter. While he becomes a powerful voice for the public effort to sequence the human genome, he recognized early on that "commercial pressure was always going to be part of the picture."[57]

While many hold this commercial pressure responsible for harnessing genomics to the logics of speed and efficiency, in many ways these logics arrived earlier, at the very moment the ABI machines landed in Cambridge. As the historian of bioinformatics, Hallam Stevens, explains: "Computers are tools of business, and they demand and enforce the kinds of practices that have transformed biological work over the last two decades, reorienting it toward speed, volume, productivity, accounting and efficiency."[58] In *The Common Thread*, Sulston describes the transformation that occurred as computers moved into the heart of his genome sequencing efforts, shaping the very design of the new Sanger Centre and bringing with them the ethos, tempos, and practices of a business. While Sulston accuses Venter of no longer being in science, but in business, just a few pages on in his account he explains that he was in business too: "The sequencers were no longer running traditionally structured labs, with a group of more or less independent scientists and a few technicians in support; we were effectively running 'businesses.' Bob [Waterston] and I had the biggest businesses at the time; Eric Lander aspired to have the biggest business."[59] Like those in the corporate world, Venter, Lander, Waterston, and Sulston sought to increase the speed and efficiency of their operations. To do this, they required ever-greater resources and tightly controlled management. Not everyone could take part. Indeed, many had to be excluded. Sulston recalled: "It looked as if we were just being pig-headed and defending our interests, but we were in a position of responsibility in more ways than one: without us the human genome would be privatized. I tried to explain this to Mike Smith, but I don't think he understood. 'You're not leaving anything for anybody else,' he protested. But if the project had been left to all the small groups, Craig would have just walked all over them."[60] At the end of the Human Genome Project, of the dozens of labs that aspired to take part, and of the twenty listed as authors on the *Nature* paper, only two, Sulston argued, had "the high level of industrial organization needed to accelerate the production of sequence" and create the final sequence: the Sanger Centre that he directed and Bob Waterston's lab at Washington University in Saint Louis.[61]

Despite his allegiance to Sydney Brenner and the ethos and practices of communal work he attributed to the LMB, Sulston found himself a central character in the transformation of biology into an industrial-scale production system that excluded all who could not keep up.[62] It was, along with other domains of knowledge, becoming a part of informatic capitalism.

Genomics and the Rise of Information as a Global Commodity

Knowledge, Jean-François Lyotard argued a decade before the launch of the Human Genome Project, had become "an informational commodity indispensable to productive power." Indeed, he asserted, it was "a major—perhaps *the* major—stake in the worldwide competition for power."[63] Certainly by the mid-1990s actors central to the HGP acted as if this were the case. Upon founding Human Genome Sciences, Craig Venter and Wally Steinberg justified their entrée into the race to sequence the human genome as nothing less than an effort to "save America's biotech industry."[64] By 2000, the fortunes of the US stock market hinged on events in human genomics. In March 2000, when UK prime minister Tony Blair and US president Bill Clinton made a public statement affirming the Bermuda principle that genomic data would be made freely available, the Nasdaq—the index of high technology stocks—lost 200 points. Thirty billion dollars were lost from the value of biotechnology stocks in one day alone.[65]

While predicted by a social theorist a decade prior, Sulston, a biologist, lived through and described these changes. "Biology," he observed, "had undergone an economic sea change—it now held the promise not only of tremendous knowledge and great benefits to humankind but also fabulous wealth. As biologists we had lost our innocence."[66] The changes altered all dimensions of the scientific life Sulston had known: its buildings, its practices, even the core value of openness itself. At the end of the Human Genome Project, the norm of openness motivated the race to finish the human genome, which in turn launched the sequencing machines "arms race." By fall of 1998, Sulston recalls, "the allies in the publicly funded project were tightening their organization to turn the tide against the threat of the invader."[67] The Sanger Centre bought thirty of the new ABI capillary sequencing machines at $300,000 apiece. Eric Lander at the Broad Institute bought 125. In 1999, the year after Venter launched Celera, ABI sold a billion dollars' worth of sequencing machines. If there was any clear winner in the race to complete the human genome sequence, it was this manufacturer of the machines. As in other domains, in the name of democratic access and inclusion, the expansion of the power to produce genomic information required concentrated wealth, and located the power to produce this information in the labs of a few.[68]

Sulston and others at the Sanger Centre did try to resist this takeover of genomics by the logics and practices of informatic capitalism. Tim Hubbard, then head of sequence analysis at Sanger, explored the possibility of using a "copyleft" agreement developed by the free software movement to protect the public project's human genome data. Such an agreement would have provided a formal legal meaning to the Bermuda principle of open access, specifying that all were free to use HGP genome data but could place no restrictions (or example, patents) on its further development. The Wellcome Trust devoted serious resources into developing this idea. John Stewart, its head of legal matters, even created a draft license agreement. However, reportedly those who oversaw the public genome databases strongly objected. The data, they argued, should remain free for all to use in whatever way they saw fit, including patenting and licensing further development and redistribution of the data.[69]

There was indeed no going back, not even for the principle of openness. The power to sequence—and thus to play a major role in the genomic revolution—already had concentrated in a few institutions. Inequalities between researchers had become institutionalized and were the price paid for universal access to the sequence of the human genome. Few traces remained of the qualities of living and learning that were so central to Sulston's early experiences of molecular biology and its moral economy of sharing.[70] Openness no longer primarily served the ethos of these communities of academic scientists. Importantly, it also fostered the needs of informatic capitalism. While the power of unprecedented flows of capital into the life sciences allured many, including Sulston, the road back from intensive flows of capital and human genome sequence to flows of fundamental knowledge remained unclear.

Genomic Information and the Problem of Knowledge

This is no small part because the road between genomic information and biological knowledge also had become unclear. The belief that information is a central good needed for right judgment and rational decision making is a core tenet of Western liberal democracies.[71] Its centrality to knowledge is also widely assumed. Consider Sulston's explanation of the importance of the human genome: it is, he explains, "the biological information itself," "the hieroglyph of biology."[72] Indeed, one way to understand why Sulston tolerated the sacrifice of long-cherished communal practices and principles to become a part of the technology-fueled ruthless race to sequence the human genome is because he believed in the profound importance of the human genome. Access to it, he argued, was of utmost importance—as important to a biologist, Sulston argued, as a dictionary is to a writer. He endured the erosion of almost all aspects of the moral economy that formed him as a scientist because he believed in the extraordinary value of the human genome and its importance for all of humanity. It was his mission to make it "public" and "free" for anyone to use—even if it meant fueling the ABI sequencing monopoly and the rise of industrial approaches to biology.[73]

Yet despite this celebration of the power and import of the human genome sequence—the biological information of life—rendering it meaningful in the postgenomic era proved anything but easy. As we will see later in this book, the problem of the meaning of genomic information came to public prominence when personal genomics companies first opened their doors in 2007, and many government officials and industry watchdogs questioned the value of genomic information for consumers.[74] Yet prior to this, in the years just following the sequencing of the human genome, a more troubling question about the value of genomic information arose: did it have value for biological and medical scientists? Some prominent genome scientists, including none other than Craig Venter, answered "No." "We have learned nothing from the human genome," Venter told the German newspaper *Der Spiegel* in 2010.[75]

Given genome scientists' promotion of the human genome as the secret of life, Venter's assertion is startling. While the suggestion that genomic information does not have value for consumers can be understood as a matter of timing—it is just too soon to know what the relevance of the basic genomic findings will be for patients—to question its meaning for biology challenges our understanding of genomic information itself. Over the course of the decade after the sequencing of the human genome, genomics would trouble the meaning and value of liberalism's most cherished principles, beginning here with the value of the free flow of information.

Genomics did not create this trouble de novo. Instead, the historical roots of genomics' problem of information dates back to World War II and the rise of cybernetics. It is at this time that information cybernetics pioneer Claude Shannon forged what cultural theorist Tiziana Terranova describes as a "technical" understanding of information that departed from the everyday "modern concept." Writes Shannon in his pathbreaking 1948 article, "A Mathematical Theory of Information": "The fundamental problem of communication is that of reproducing at one point either exactly or approximately a message selected at another point. . . . Frequently the messages have meaning; that is they refer to or are correlated according to some system with certain physical or conceptual entities. These semantic aspects of communication are irrelevant to the engineering problem."[76] In this, Shannon's theorization of information, information is a message (or signal). The problem is how to relay it in a manner that minimizes its distortion due to noise. Importantly, this problem does not address the problem of what the message means. The task is only to *encode*—not decode—the signal in a manner that secures its accurate reproduction. In other words, information is a signal to noise problem. Preserving the integrity of the signal is tantamount. In the context of a world in which military operations and communication increasingly happened across nations and seas, accurate transmission of a message could be the difference between victory and defeat, life and death.

Terranova argues that today the "modern concept of information" as meaningful content has been "explicitly subordinated to [these] technical demands of communication engineering."[77] What matters is the quick, efficient, and faithful transmission of signals. This is what fuels the massive growth of the information economy. Twitter does not primarily care what the tweets you send mean, rather that you send and receive them.[78] Media theorist Jodi Dean describes this demand to constantly send and receive signals as the central feature of communicative capitalism.[79] The meaning of these signals, Terranova and Dean both argue, is largely irrelevant in the informational milieu of the twenty-first century.[80]

Popular accounts of genomics challenge this understanding of our contemporary informational world and embrace instead the modern concept of information as meaningful content. Decoding, not just encoding life, is the stated goal. Venter's own autobiography announces this in its title, *A Life Decoded: My Genome, My Life*. Inside the covers he refers to "the human genome" as "the book of life" that reveals "the secrets of our inheritance." Throughout, he turns to his genomic information to make sense of everything from why he has asthma to why he prefers to work at night.[81] Francis Collins, head of the public effort to sequence the human genome, promises something arguably even more profound in his own autobiographical account, *The Language of God: A Scientist Presents Evidence for Belief*. Collins, paraphrasing former US president Bill Clinton, describes the human genome as "God's instruction book."[82]

However, a closer read of these accounts combined with an attention to genome scientists' everyday practices, reveals a more fraught relationship between genomic information and meaning. Consider that to render life loquacious—to make it communicate its As, Cs, Gs, and Ts—genome scientists inscribed it in silico. They abandoned biology's analog approach to sequencing—the Sanger technique—for a digital one. The former entailed what Collins described as the "arduous task" of pouring agarose gels between two glass plates, hand-preparing radioactive DNA samples, hand-loading the gel, laying the resulting gel on film, and reading the resulting X-ray-induced images. Both Collins and Venter argue—and I concur based on my own experience pouring and reading these gels—that this process is prone to many problems, including that of human interpretation. Lanes do not run parallel. The further down the gel one reads, the less distinct the black marks and gaps become. Is that a T or a C, and A or a G? Such questions occupied my mind and those of the many others who spent hours poring over films of gels, trying to record correct sequence.

For Venter, the awkward, slow nature of humans interpreting the meaning of a hash mark on a gel was the central rate-limiting step in genomics. Indeed, he believed that all human interpretation impeded progress: "I found the room for interpretation particularly frustrating because I had such high hopes for molecular biology. Too many times I had seen science driven less by data and more by the force of a particular personality or the story on which a professor had built his career. I wanted the *real*, empirical facts of life, not those filtered through the eyes of anyone else."[83] And so, Venter claims, he set out to replace humans with machines and implemented for the first time "automatic" sequencing on a large scale. Describing the ABI sequencing machines that made this automatism possible, Venter writes: "The four colors, representing the different nucleotides, provided a *direct readout* of the genetic code, transforming the analog world of biology into the digital world of the microchip."[84] The ABI machines, Venter believed, eliminated human-filtered readings of DNA. Electronic computers directly encoded base pairs as clear 0s and 1s.[85]

The public effort quickly followed suit, and also invested in digital automation. As in other realms of human inquiry, this move took biology out of the data desert of the analog age and into the data deluge of the digital one. Fast, efficient, computer-operated machines that shined light beams onto microcapillaries and recorded results on silicon chips replaced humans who poured and processed large unwieldy gels. Venter proudly describes how his automated machines "eliminated the need for thousands of workers."[86] The digital overthrew flesh.

For Venter, it also overthrew pesky government bureaucrats who worked at the NIH, and who on his account failed to support his research due to petty politics. Venter no longer needed them. The tremendous growth of the digital economy created venture capital for anything that promised to increase the demand and flow of bits and bytes. Human genomics, with its need to encode billions and billions of nucleotides, promised this and more.

Leaving people behind, Venter "rocket[ed] forward" into the genome age.[87] With the vast sequencing and computational facility he built at The Institute of Genome Research (TIGR)—what he described as "the beating heart of the organization"—he produced more genomic data than anyone else.[88] But once rocketed off mortal earth—plagued as it is by the slow and painful stuff of human interpretation—did he also leave the land of meaning?[89]

Throughout his memoir Venter repeatedly stresses his commitment to interpreting the meaning of genetic sequences. As a medic in Vietnam, and then a biochemist studying adrenaline receptors, life and what made it tick became his central concerns. Indeed, he describes his interest in interpretation as a key difference between him and then-director of the Human Genome Project, James Watson: "Watson argued that our goal was to work out the sequence and let future generations of scientists worry about understanding it. I had always believed that interpretation was crucial to making the sequencing both efficient and worthwhile."[90] In practice, though, sequencing and speed became paramount. On Venter's own account, he ultimately left the field of receptor biology to become a genome scientist, and while he kept his interest in the biological meaning of the genetic code, he realized that for the most part he "would have to fight that battle another day."[91]

When others did try to work with Venter to fight that battle, it was never an easy task. The company created to make good on the investment in TIGR, Human Genome Sciences (HGS), faced significant problems translating genomic data into something valuable for shareholders and patients. Explains Venter himself: "Paradoxically, the torrent of data generated by TIGR, which should have been an achievement to be celebrated, was the source of the problem: HGS was simply overwhelmed. . . . Had I delivered them a single gene that was linked to a disease, they would have known how to mount a major discovery effort to turn the find into a test for new drugs. But I had given them thousands of genes over the course of a few months. . . . HGS complained that to exploit the data 'was like trying to drink from a fire hose.'"[92] As the stories of this book will bring alive, the problem of how to make sense of the massive amounts of data that genomics generated quickly became the problem faced by both corporate and public-funded genomics initiatives.

This problem of information overload was not unique to biology. Nor was it a new problem. In his 1981 book *The Political Unconscious*, the cultural theorist Fredric Jameson argued: "Unfortunately, no society has ever been quite so mystified in quite so many ways as our own, saturated as it is with messages and information, the very vehicle of mystification."[93] Theorists Jean Baudrillard and Mark Poster also have made similar observations, arguing that the increased speed and volume of informational language threatens the representational function of symbolic systems.[94] As early as the mid-1990s, writers were coining terms such as *data smog* (which made it into the Oxford English Dictionary in 2004) and warning of the problem of information overload and the eclipse of meaning. More recently, University of California, Berkeley, professor Geoffrey

Nunberg wrote in his *New York Times* book review of James Gleick's *The Information*: "There are no roads back from bits to meaning." The technical units of information—0s and 1s—just don't correspond to anything of social significance.[95] Genomics, rather than creating a brave new world that created unique challenges, fueled this existing problem of meaning and extended it into our own flesh and blood.

Forging Routes Back from Bits to Biology: The Problem of Postgenomics

In 1984, at the first Hacker's conference, the counterculture icon Stewart Brand made a claim that would go on to be a rallying cry for the digital age: "Information wants to be free." At the time, Brand's words referred to Moore's Law and the law-like manner in which the cost of producing information had decreased over time.[96] While his was a description of a physical and economic phenomenon, the phrase soon came to carry strong moral overtones: not only is information nearly free to produce, it *ought to be* freely disseminated. Animated by this sense of justice, genome scientists during the Human Genome Project era became very good at freeing digitized genomic information. In an ever-cheaper and more precise manner, they encoded the As, Cs, Gs, and T's of the flesh and faithfully relayed them to computer screens. As a result, genomic information became a lively material reality. The genomics world became abuzz with discussions of where to store it, how to build computers big enough to process it, and how to provide the electrical juice needed to keep it alive and well.[97]

Yet this was a long way away from what many promised genomics would be abuzz with: medical breakthroughs, profound new understandings of life. While born out of a concern for the senseless loss of human life—in Hiroshima, Nagasaki, and Vietnam—in practice genome scientists built worlds where it was possible to assert, as Venter did, that the machines, not the humans, had the beating hearts. While they sought the information of life, what they clearly had created was the life of information.

That act created a fundamental question: Does genomic information have value? As George Church, head of Harvard's Personal Genome Project, presciently observed: "When costs come down to zero then people have to decide how to add value to it [the genome]."[98] While Church made this comment in 2009, the problem of adding value to genomic information became the problem as soon as the Human Genome Project came to a close.[99] Key to solving this problem, Church and others realized, would not be lots of machines, but lots of people.

Yet who would those people be? Genomics was built around a problem of information that provided few obvious routes back from bits to biology, from sequencing machines to people. If genomics was to become meaningful for broad publics, this had to change. In the chapters that follow, I tell the stories of what happened as genome scientists appealed to different concepts and visions of liberal democracy as they sought to recruit human beings from around the world to give their DNA and data for genomic study.

3

INCLUSION
Can Genomics Be Antiracist?

The important conclusion of human population genetics is that races do not exist.
—Luca Cavalli-Sforza, in a speech at Tuskegee University

You can talk all day about race not having biological significance, but I am convinced it will be another 100 years—if we are lucky—before Americans will stop acting as though it has biological significance.
—Patricia King, in a speech at Tuskegee University

Leaders in human genomics realized that recruiting the human beings whose DNA and data would be needed to make genome sequence information meaningful posed a daunting problem. Many remembered all too well the controversies that surrounded the first effort to collect "diverse" human genomes, the Human Genome Diversity Project.[1] In the early 1990s, human population geneticists proposed to survey human genetic diversity through sampling globally what they described as "isolated" indigenous populations.[2] Some biological anthropologists and many indigenous rights organizations charged the proposers with treating indigenous peoples as mere objects of study, not human beings with the right to own and control their own DNA. The proponents of the Diversity Project, they argued in prominent academic and science news outlets, threatened to instigate a new form of biological racism and colonialism.[3]

The importance of these early controversies cannot be overestimated. Still today directors at the US National Human Genome Research Institute (NHGRI) recall the significance of these high-profile controversies.[4] For years afterward, NHGRI sought to steer clear of them by putting to one side questions about how human genomes differ. Instead of funding their own systematic studies, they invoked the widely cited findings of previous human population genetics research: humans are all remarkably similar at the genetic level; we are 99.9 percent the same; 85 percent of those few differences vary between individuals, not groups.[5] From 1995 to 2000, commentators in the popular and scholarly press interpreted these findings to mean that there is "no such thing as race" at the biological level.[6] With a new millennium on the horizon and a new science in the making, prominent social theorists imagined that these findings might bring about a more just society, one without race.[7]

While this hopeful story helped to quell the controversy, it also set up a dilemma. Despite the emphasis on human similarity, genome scientists believed that differences between genomes could help explain the differential manifestation of human diseases. Thus, as the decade of the 1990s came to a close and leaders of the HGP began to consider what would come after the sequencing of the human genome, most recognized the need to address the question that for years they had so astutely avoided: how do human genomes differ, and how do those differences matter?

Given the enduring legacy of slavery and the political importance of including African Americans in the US body politic, maybe it is not surprising that efforts to include them in genomics research were among NHGRI's first attempts to address this dilemma.[8] The interest in "African American" genomes also aligned with the Out-of-Africa theory.* This theory posited that the human species first emerged in Africa, and so "African" genomes, by virtue of being the oldest, contained the most human genetic variation. Studying the genomes of African Americans in the United States, some leaders at NHGRI believed, might enable studying this variation while averting charges of colonialism sparked by the Diversity Project.

Yet in other ways, the proposal to collect the genomes of African Americans went against then-dominant modes of making sense of differences between human genomes. In particular, it stood in tension with the strong moral claim that at the genomic level, one's race did not matter. Confusion resulted that did not escape NHGRI leadership. At the turn of the millennium, Francis Collins, then director of NHGRI, gave public talks in which

* I place "African American" and "African" in quotes to indicate that the meaning and proper use of these and other human group categories are not self-evident, but rather are defined in particular contexts. For example, as we will see, many would contest the idea that the category of "African American" should be used to denote particular kinds of genomes. I will place these and other human group categories in quotes when I first use them, but for ease of reading, I will not always use them upon repeat use.

he quoted Harvard historian and African American scholar Evelyn Brooks Higginbotham: "When we talk about the concept of race, most people believe they know it when they see it, but arrive at nothing short of confusion when pressed to define it."[9] What is race? What would it mean to say it does not exist? In the realm of human genomics, not only did these questions create confusion, they also raised questions about the constitution of a just, democratic society. Would collecting DNA from African Americans for genomic analysis advance or hinder the cause of justice?

As one century ended and another dawned, I traveled to many of the sites where these questions were asked and debated: the National Human Genome Research Institute and historically black colleges and universities (HBCUs) in the United States that sought to establish human genome research initiatives.[10] At each site I met and interviewed genetic researchers, ethicists, health care professionals, members of institutional review boards, and bioethicists.[11] All expressed deep commitments to advance the goals of antiracism and human understanding. Yet not all agreed on the path to justice and knowledge in the realm of genomics.[12] Is it possible to create a just, antiracist form of genomics that includes African Americans in the promised genomic medicine of the future while basic health care is denied to many African Americans in the present?[13] As scientific and political leaders prepared to celebrate the completion of the draft of the human genome sequence in June 2000, natural and social scientists, community leaders, community members, genetic counselors, and policy makers on HBCU campuses grappled with this question. While no clear answer emerged, the debates revealed some of the central issues and tensions that would define the decade ahead.

Can Genomics Be Antiracist?

"Scientists Say Race Has No Biological Basis," read the front page of the *Los Angeles Times* on February 20, 1995.[14] "Don't Classify by Race, Urge Scientists," reported the *Boston Globe* that same year.[15] These and other prominent articles in the popular press reported the then-prevailing views of leading biological anthropologists and population geneticists. Prominent cultural theorist and public intellectual Paul Gilroy invoked their arguments in his 2000 book, *Against Race*. While at the beginning of the century the prominent African American sociologist W. E. B. Du Bois argued that "the color line" would be the problem of the twentieth century, at the century's close Gilroy—also a scholar of African descent—believed that scientific research announced its demise.[16]

By 1995, this had become the dominant story about race and genomics told across racial lines. However, it was not the only one.

For years, a group of African American immunologists, geneticists, and biological anthropologists had been arguing for the need to study the biology of African Americans. They pointed to studies that revealed that African Americans suffered disproportionately from high rates of diabetes and hypertension, both of which can ultimately cause kidney failure.[17] While renal failure can be treated through kidney transplants, for a transplant to be successful, the human leukocyte antigen (HLA) type of the new kidney must "match" the donor.[18] In the early 1980s, when human geneticist Georgia Dunston began her HLA work at Howard University, a historically black university in Washington, DC, HLA-D type matches failed 44 percent of the time; in "Caucasians," the failure rate dropped to only 2 percent.[19] The reason, she argued, was that HLA antisera used for the matching reagents came primarily from "multiparous Caucasian women."[20] These cells preferentially identified HLA types for "Caucasians." Dunston set out to address this disparity by collecting HLA antisera from multiparous African American women recruited largely from Howard University hospital patients.[21]

In addition to these health benefits, she and other African American researchers argued there was another important reason to study this molecular diversity: it could demonstrate the African roots of all humanity. The Out-of-Africa theory first put forward by Charles Darwin in the *Descent of Man*, and brought to prominence by mitochondrial DNA researchers in the 1980s, suggested that "African" genomes represented the widest range of diversity in the human genome. This was so, or so the theory posited, because the human species first evolved in Africa and so the genomes of its inhabitants had the most amount of time to accumulate differences.[22] Thus, if you wanted to create an inclusive map of diversity in human genomes, these genomes should act as the referent.[23]

Yet while African American geneticists and anthropologists believed this argument to be scientifically sound, they worried the idea was not politically viable: "Do you think America is ready to put major research dollars in characterizing a set of samples from African Americans to be the reference for humanity?"[24]

Researchers at Howard University sought to find out. In 1990 Georgia Dunston submitted a proposal requesting additional funding for the Immunogenetics Laboratory at Howard to collect DNA from African American families. Researchers and funders intended these samples to supplement those that had been collected by the Centre d'Étude du Polymorphisme Humain (CEPH) in Paris, France, and that had been used by HGP researchers to create the first genetic maps.[25] CEPH's samples came mostly from Mormon families from Utah. Howard University researchers argued that this biased the endeavor, preventing it from serving large portions of humanity, including many African Americans.[26] Given the importance of the Human Genome Project, the largest and costliest science project since the Apollo space mission, prominent African American researchers found this state of affairs unacceptable. Fatimah Jackson, a biological anthropologist and an advisor to Howard University's efforts argued: "If, as Eric Lander suggests (1996), the HGP will provide the 'structural periodic table' from which functional genomics will emerge, we must ask, 'will the variation most frequently encountered among African Americans be reflected in this molecular periodic table?'"[27] To prevent the HGP from becoming another example of accepting "European-American (and male) patterns as human universals," she urged "full inclusion" of African Americans.[28]

NIH awarded the Immunogenetics Laboratory at Howard University funds to develop this initiative.[29] Planning of what became known as Genomic Research in African American Pedigrees (or G-RAP) took place from 1990 to 1993, during which time Howard researchers visited the CEPH in Paris to learn from its leaders about their record keeping and management procedures. G-RAP leaders also held planning meetings to answer a critical question: Which African American families should be selected to represent African genetic diversity?

In so doing, they challenged the views of many population geneticists who at the time believed that the difficulties of historical reconstruction combined with "racial admixture" prevented defining African Americans for the purposes of understanding human evolution. Building on Alex Haley's work in *Roots: The Saga of an American Family*, they argued that it was possible to reconstruct the history and genealogies of African Americans through looking at the records of slave ships.[30] While slavery might have torn families apart and obscured historical connections, it also was a big business that kept detailed records of where slaves originated, when they were purchased, and by whom.[31] Leaders of G-RAP believed that this information could be used to reconstruct the history and demography of African Americans.[32]

Investing in this effort, they argued, would not just benefit African Americans, it also would contribute to genomics' broader exploration of what it means to be human. As one G-RAP organizer explained: "Everybody knows what the human genome is. But then, what does it mean? Who is human? What is it that defines us humans?"[33] Mapping diversity in the genomes of the peoples of the African continent, organizers of G-RAP argued, would ensure that genome scientists' answers to these questions represented the greatest range of human diversity. It also promised a positive story about racial difference. This possibility spoke powerfully to one leader of the initiative: "I was always intrigued with what you might say is the philosophical level of the question. There must be some good reason for why we are different. Obviously, it can't be the reasons that seem to be those of society because the things that I characterized myself as in society were bad. My color was bad, my hair was bad, my physical features were not preferred."[34] G-RAP researchers hoped that genomics could reverse this painful story through revealing the good in human diversity: "I've always thought that this whole initiative [the Human Genome Project] had a larger agenda for humankind that has not been fully recognized . . . and that's what we have to learn of ourselves through the mirror that differences make it possible for us to see. . . . The more I come to know who I am, the more I realize I'm one with the whole."[35] For leaders of G-RAP, this was the potential power of African American inclusion in human genomics. It would not provide just access to a broader range of health care. More profoundly, it would take society "to the next level," where unity, not division, prevailed.[36]

Contesting the Antiracist Potential of Genomics

However, all did not agree. In 1990, the prominent African American scholar Troy Duster warned in his *Backdoor to Eugenics* of the dangers of turning to genetics to understand an ever-widening array of social phenomena: from criminality to drug abuse to alcoholism to rape.[37] Rather than address underlying social changes—such as the widening gap between rich and poor—Duster observed that policy makers increasingly favored explanations that framed some people as "bad seeds."[38] These biological explanations justified Reagan-era cuts to social services.[39] A backdoor to eugenics, Duster argued, had opened.

Duster's book was widely reviewed in prominent science and medicine journals such as the *New England Journal of Medicine* and the *American Journal of Human Genetics*. A prominent sociologist, grandson of suffragist and civil rights leader Ida B. Wells, and member of several National Institutes of Health review panels, Duster's concerns did not go unnoticed. Shortly after the publication of *Backdoor to Eugenics*, HGP leaders asked him to testify before the Working Group on Ethical, Legal and Social Issues at the National Center for Human Genome Research (NCHGR, the precursor of the National Human Genome Research Institute), and in 1993 he joined the group as a member.

While Duster might have sounded the alarm, critics of the Human Genome Diversity Project turned these concerns into a pressing reality. While some—including proponents of G-RAP—viewed the Diversity Project as an important corrective to the "Eurocentric" view of the genome provided by the HGP, many others feared a new wave of scientific racism.[40] Critics dubbed the proposed initiative the Vampire Project, a project more interested in sucking the blood of indigenous peoples than in their ultimate survival.[41] Indigenous groups and indigenous rights organizations around the world condemned it.[42] Prominent journals, including *Science*, *New Scientist*, and *Nature*, kept the controversy in the spotlight.[43]

The protest shocked many. Like G-RAP, the proposers of the Diversity Project celebrated the antiracist potential of studying human genetic diversity.[44] Indeed, I myself almost went to graduate school at the University of California, Berkeley, in order to work on the Human Genome Diversity Project because I believed in its proposers' antiracist humanistic vision. However, for many indigenous groups, as well as critical biological anthropologists, the project represented another example of white European scientists framing indigenous peoples as "primitive" others and treating them as mere objects of research.[45] Far from antiracist, they feared the initiative would revive scientific racism. As I experienced over the years that I followed and wrote about the Diversity Project, these divergent views left their scars on scientists and indigenous rights leaders alike.

NIH subsequently steered clear of any research on human genetic diversity for several years.[46] In February 1994, a population geneticist interested in coming to NIH was told that NIH had no interest in human genetic diversity research: "I was talking to people here at the genome center in early 1994 and people here were saying 'You know, we don't really see much interest in genetic variation. What's it important for?' So that's a date, February '94: No interest in genetic variation."[47]

Not only did NIH have no interest in a scientific study of genetic variation, they also avoided ethical analysis of it. This became a point of contestation in the mid-1990s, after the publication of *The Bell Curve*.[48] Duster and other prominent social science commentators worried that this book would be used to justify the further erosion of social support for underserved minorities.[49] Members of the ELSI Working Group at the NCHGR on which Duster sat asked the American Society of Human Genetics to critically assess its claims about the links between race, IQ, and genetics. However, Francis Collins, then head of the NCHGR (now director of the NIH), believed no response should be issued. Collins and the American Society of Human Genetics argued that *The Bell Curve* discussed behavioral genetics, an area that they as molecular geneticists did not engage.[50] Duster responded that it would not be long before molecular geneticists began to make claims about traits like intelligence.[51] Time would prove him right. Yet at the time, as Duster recalled: "Collins and that group didn't buy it. They locked in saying that was not their purview . . . Collins wanted to shut [the discussion] down. We said no, and there was a huge conflict."[52] While Collins did prove an ally in opposing Myriad Genetics' breast cancer gene patents, no one in the human genomics community wanted to address *The Bell Curve*.

G-RAP researchers sought to move their initiative forward in the context of this effort to distance human genomics from issues of race and racism. In 1990, when they submitted their planning grant, it was possible to argue that including African American genomes in the creation of the referent map of the human genome would act as an antiracist corrective to the HGP. By 1994, when they presented a full grant to NCHGR, understandings of the human genome had shifted. The fragile consensus that in the eyes of the human genome we were all one had come under scrutiny. Funding an area of research that might inadvertently further trouble this consensus was not forthcoming. NCHGR turned down the G-RAP request for further funding. Bettie Graham, a program officer at NCHGR, offered the following explanation: "The plan was not what we had envisioned; it was for a much larger project than we had anticipated."[53]

The planning process had indeed expanded G-RAP's vision, bringing questions about the scientific and ethical basis of representing African Americans into focus. Reportedly, these charged issues lay outside NCHGR's mandate.[54] In an environment where links to scientific racism could lead to the ends of careers and projects, NIH decided it was prudent to draw lines: genomics innovated new tools and methods for mapping and sequencing genomes; anthropology addressed issues of how to define and represent human genetic diversity.

Under the Threat of Patents, One Vision of Genomics and Race Moves Forward

This turn away from diverse humans, and back to tools and machines, did not last long. On July 28, 1997, Abbott Laboratories announced a deal with the genomics company Genset of Paris: Abbott would invest $20 million of equity in Genset over two years and support up to $22.5 millions of Genset research in return for access to the company's genetic variant data and rights to market it to other companies. Through agreements like this, many worried that the study of the human genome in all its variation might become big business, walling out big government.

NHGRI responded quickly. One week after the announcement of the Abbott and Genset deal, at a retreat of all leaders of Institutes at the NIH, Collins sought support to create a publicly available map of single nucleotide polymorphisms, dubbed "the SNP map." One month later, on September 11, 1997, the NHGRI advisory council agreed to launch a public-private partnership (what became known as the SNP consortium) to support the rapid creation of an SNP map. As one academic scientist involved in the discussion explained: "Francis Collins was in a race with commercial outfits. . . . As I understand from what Collins said at these meetings . . . private industry said 'Ok, if you do it, we won't do it, and then it will be public domain variance. But if you don't do it, we are going to do it.' Now it may not have been so amicable as the way it was presented, but he was afraid [that SNPs would be locked up by patents]."[55] While addressing one ethical problem (privatization of genomic data), the plan to move rapidly into human genetic variation research raised another: the appropriate source of diverse human DNA samples on which to conduct SNP discovery.[56]

Pressure to study the genomes of diverse human groups had reached a crescendo. Researchers at Howard were no longer the only ones pushing for inclusion of genomes specifically identified by race and ethnicity; the broader biomedical and genomics community joined them, and changes at the policy level demanded it.[57] In June 1993, Congress passed the National Institutes of Health Revitalization Act, a law that required minority inclusion in medical research.[58] By the mid-1990s, researchers began to feel its effects in research grant applications that now required them to include plans for minority inclusion. Yet despite a clear call for inclusion from both the research and policy communities, how this should proceed remained far from clear. No place would demonstrate this more than Tuskegee, Alabama.

Can Genomics Transform the Legacy of Racist Science?

From 1932 to 1972, the US Public Health Service (PHS) studied approximately four hundred African American men from in and around Tuskegee, Alabama, who were presumed to have late-stage syphilis.[59] After some initial treatment, the PHS stopped treatment of the men's syphilis, giving them instead aspirin and iron tonic. The men were told these were treatments for their "bad blood." Instead, they were enticements to keep them coming back so that the PHS researchers could study the natural course of syphilis by doing spinal taps, drawing blood, and performing autopsies on the men over time. The men were never told they had syphilis. They were not given penicillin, even after this became the standard of care in the late 1940s. The story of the study first broke in the *Washington Star* on July 25, 1972. A day later it appeared on the front page of the *New York Times*. In the decades since, the name Tuskegee became synonymous with scientific racism and abuse of human subjects. How then did it become the site for rekindling the discussion about genomics and race in the late 1990s? After years of distancing genomics from associations with past instances of scientific racism, why return to the "scene of the crime"?[60]

The answer lies partially in the broader reworking of the meaning of inclusion in biomedical research that took place between the mid-1980s and the early 1990s. Prior to this period, few paid attention to who was included in research unless abuses came to light. However, as Steven Epstein accounts in his book *Inclusion: The Politics of Difference in Medical Research*, during the 1980s a new wave of women's health advocates and members of the Congressional Caucus for Women's Issues began to argue that the underrepresentation of women in biomedical studies presented a critical problem: it led to a low profile for women's health issues and slow progress in the advancement of knowledge about conditions that disproportionately affected women, most notably breast cancer.[61] Members of the Congressional Black Caucus also gave their support. In 1993, this alignment of interests led to the addition of a section in the NIH Revitalization Act that made inclusion of women and minorities in NIH-funded research the law of the land.

Yet while the meaning of inclusion in biomedical research changed at the national policy level—transforming it from a problem or nonissue to "a domain in which social justice could be pursued"—on the ground biomedical researchers found a different reality.[62] Efforts to recruit minorities into research often met with nonresponse or refusal. The legacy and lore of the Tuskegee syphilis study remained powerful. Indeed, in 1994 it contributed to the derailment of President Bill Clinton's effort to appoint Henry W. Foster Jr. as surgeon general. Employment at Tuskegee during the time of the study proved enough to tarnish Foster's image.[63]

Whether for these or other reasons, by the mid-1990s attention refocused once again on Tuskegee. In February 1994, at a meeting titled "Doing Bad in the Name of Good: The Tuskegee Syphilis Experiment and Its Legacy," University of Virginia bioethicist and former Episcopalian priest John Fletcher called for an apology for the study. As medical historian Susan Reverby explains in her groundbreaking book about the study and its legacy, Fletcher worked his considerable connections at the NIH and the US Centers for Disease Control (CDC) to create support for an apology. In May 1995, a working group at the CDC formed.[64] In January 1996, historians, federal health officials, leaders of Tuskegee University, and state and local health administrators gathered at the Tuskegee Kellogg Conference Center to explore the impact of the syphilis experiment on minority participation in research, and to develop a strategy for moving forward with an apology.[65] In May 1996, a report drafted by Fletcher and reviewed by other members of the meeting made its way to CDC. The report called for both an apology and a bioethics center and museum to be established at Tuskegee that would "transform the legacy."[66] Tuskegee, the report envisioned, would no longer be merely a metaphor for racist abuse, nor would it just be repaired; instead, its students and faculty would become world leaders in the field of bioethics.[67]

These efforts to rework the meaning of both Tuskegee and studies of human genetic variability—from emblem of racist past to site of antiracist future—coincided. Thus, when Ed Smith, a turkey geneticist and young Tuskegee faculty member, contacted the Ethical, Legal, and Social Implications programs at both the Department of Energy and NHGRI in early 1996 to see if they would support a meeting on the Human Genome Project at Tuskegee, he received an enthusiastic response. Smith, like Dunston and Jackson before him, believed that if the Human Genome Project was to be the next Apollo project, a "holy grail" of both great economic and symbolic importance, then African Americans should take part. They could not wait for the benefits of the project to "trickle down." Instead, they required institutional support to facilitate human genomics research by and for African American people.[68]

Unlike Dunston and Jackson, Smith's timing could not have been better. Leaders of the HGP realized that support of African American researchers might prove important to their efforts to launch human genetic variation research. Hosting a meeting at a historically black university might be an important first step. It would demonstrate some interest and support for human genomics among African Americans and spark the dialogue needed to further transform the image of the field in African American communities.

But why Tuskegee? Would not hosting a meeting on human genomics at the site of the syphilis study rekindle links between genomics and scientific racism? Meeting organizers argued "no." Instead, they reasoned, Tuskegee was a particularly powerful place in which to demonstrate that genomics supported antiracist ends. Leaders of the university, in turn, hoped that bringing an important discussion of genomics to its campus would help transform Tuskegee's image from site of the most infamous example of scientific racism to thought leader in integrating sciences and ethics. Thus, with great support from Tuskegee, DOE, NHGRI, and the US Department of Agriculture, a meeting titled "Plain Talk about the Human Genome Project" took place at the Kellogg Conference Center on the Tuskegee University campus from September 26 to 28, 1996.

Plain Talk?

Speakers at the meeting included luminaries from the world of human genomics and genetics: David Botstein, then chairman of the human genetics department of Stanford University; Richard Myers, director of the Stanford Human Genomics Center; Maynard Olson, University of Washington (UW) genome scientist and developer of yeast artificial chromosomes (YACs) that enabled study of large chunks of genomes; and Luca Cavalli-Sforza, a founding father of human population genetics. All participants gave talks that promoted the benefits and antiracist potential of the Human Genome Project. Luca Cavalli-Sforza stated simply: "The important conclusion of human population genetics is that races do not exist."[69] David Botstein celebrated the successes of the Human Genome Project while discounting fears that genes could be linked to traits as complex as IQ, and thus provide a basis for discrimination.

However, not all the speakers agreed. Patricia King, Georgetown University law professor and member of the national commission that drafted the Belmont Report, observed: "You can talk all day about race not having biological significance, but I am convinced it will be another one hundred years—if we are lucky—before Americans will stop acting as though it has biological significance. Until we understand that, my view is that we are nowhere."[70] Troy Duster, also a speaker at the conference, began his talk with a joke about three African American students who came out of their university classes in which they learned that race had no meaning only to find that they could not hail a taxi. Both Duster's and King's points were clear: unless we recognize how claims about genetic difference can be used to discriminate in a society that continues to be racially stratified, we remain unable to understand and address the potential harms of genomic studies of human differences. These arguments resonated with the audience.[71]

The nature of the discussion and the lines of the debate roughly paralleled those that took place in the early 1990s when Georgia Dunston and Fatimah Jackson—also speakers at the Tuskegee meeting—launched G-RAP: some argued that genomics would help undermine race and thus racism; others argued that until underlying social structural inequalities were addressed it was naive to believe that genomics had this antiracist potential. There was, however, one critical difference. This time, Human Genome Project leaders keenly listened to the debate and sought an alignment between African American communities and the HGP that would address concerns and offer benefits. One possible way to do this emerged at the meeting: establish a genome center at Tuskegee.

Maynard Olson put the idea forward in his talk titled "Genome Centers: What Is Their Role?" Olson, widely respected in the genomics community as an intellectual guru and technical master, reasoned that since the technical problems of sequencing a human genome had been solved, it now was both possible and desirable to open up genome sequencing to a broader group of scientists.[72] Although this would somewhat increase the final cost of the Human Genome Project, he argued it would bring broad benefits.[73] Most importantly, it would help bring diverse people into genomics. For Tuskegee, establishing a genome center promised new resources and an opportunity to rework its identity from one tied to the atrocities of the past to one propelled by promises for the future.

An alignment formed. President Clinton apologized for the Tuskegee Syphilis Study and plans commenced to establish a "satellite" genome center at Tuskegee all in the same year. On May 16, 1997, Clinton offered his apology from the East Room of the White House.[74] In October 1997, in a postscript to his "Plain Talk" presentation, Olson reported Tuskegee's move forward into genomics. A "satellite facility" for human genome sequencing at TU, he wrote, would be established with "advanced instrumentation for carrying out four-color fluorescence sequencing."[75] In retrospect, the descriptor "satellite" should have flagged the problems to come.

Yet it is easy to see how at the time it seemed a perfect fit. Leaders at NIH felt pressure to create resources that would make it possible to study human genetic variation. Dunston and Jackson had been arguing for years that the US government should provide resources to study genomes of African origins. They believed that sampling in the crescent states of the United States—where slave traders shipped the enslaved, and where many of their descendants remained—made the most sense. Alabama was a crescent state, and Tuskegee itself was located in the heart of the Black Belt.[76]

Studying the genomes of African Americans in the Black Belt also converged with the emerging scientific interest in mapping human genetic variation. One of the first NHGRI funded projects that Ed Smith and Maynard Olson pursued—"Linkage Disequilibrium Analysis of the DARC Gene Containing Region of the Short Arm of Human Chromosome 1"—proposed sampling African Americans in the Black Belt for the explicit purpose of understanding the structure of human genetic variation, knowledge critical to assessing the feasibility of different mapping strategies.[77]

In addition to advancing NHGRI's interest in understanding patterns of human genetic variability, studying the Duffy Antigen/Chemokine Receptor (DARC) locus advanced the desires of some African American scientists to characterize African Americans at the genomic level. Because of the protection a variant of this locus offered to malaria, it was thought to occur in almost 100 percent of people of African origins and almost 0 percent of whites.[78] Thus geneticists posited that it could measure levels of "admixture" of African and non-African genomes, a measure they believed important to efforts to sort variation due to disease from variation due to what they called evolutionary history.

From Talk to Practice: Poverty and Inequality Meet High-Tech Science

Yet while scientific interest and support appeared in sufficient alignment, the ethical grounds for the project proved less clear. As Smith and Olson soon discovered, any effort to collect DNA samples in the Black Belt required navigating long-standing contentious questions about how to approach and recruit research subjects, and how to responsibly use any data or samples collected from them. It should not have surprised anyone that doing this at Tuskegee might raise particular concerns. However, Olson's postscript written in October 1997 makes no mention of these ethical complexities. Instead, his tone is matter of fact: "Upon his return to Tuskegee for the 1997–1998 school year [Smith] will begin collecting human DNA sequence at the University."[79]

Indeed, when he returned Smith did begin collecting blood for DNA extraction from approximately two hundred African Americans in and around Tuskegee. He employed an African American medical technologist to oversee the blood draw. The medical technologist placed advertisements in the local paper to get the word out about the research. Collections took place at county health fairs. In order to take part in the study, all participants signed an informed consent form. Yet while following all the proper and accepted procedures, those close to the study worried that so many expressed interest in taking part not because they wanted to participate in scientific research but because "we have twenty-five bucks attached to it."[80] The US Census put the poverty rate in Macon County at 32.8 percent in 1999, placing it in the top ten poorest counties in the nation.[81] The possibility that people gave blood because they needed the money, and not because they understood and supported the research, seemed high. Ultimately, the researchers involved with the project decided the ethical challenges were too great to use the samples.[82]

Lack of resources for the science also challenged the viability of the research. University of Chicago researchers interested in DARC already had presented their research at conferences on the way to a 2000 publication in the *American Society of Human Genetics*.[83] While both Smith and Olson and the leaders at NHGRI who supported their work envisioned leveling the playing field so that a more diverse group of scientists could enter genomics, genomics had become big science, requiring large amounts of economic, intellectual, and ethical resources. Entering the game required more than the efforts of a few individual researchers and a funding stream designated to bringing underrepresented groups into genomics.

To address this problem, Smith and his colleagues sought to bring a major NIH grant to the campus that would provide funds for research as well as the overhead dollars needed to build the institutional resources required to compete in the high stakes field of genomics. They sought a research topic for which TU would bring a competitive advantage and one of interest to researchers at the University of Washington so that UW would continue to provide training support. Finally, given the legacy of the syphilis experiment, they sought a research project that would bring clear benefits to the local communities.

That project would be a study titled "Apolipoprotein B Variation and Coronary Heart Disease Risk in African Americans."[84] Deborah Nickerson, another leading genome scientist at UW, studied variability in the *ApoB* gene, a gene involved in lipid metabolism. Tuskegee University researchers studied the effect of diet on cardiovascular disease risk factors at the TU Center for Research on Diet, Lifestyle, and Cardiovascular Disease in Rural Alabamans. The plan was to join up the UW genomic research with the TU environmental research to create an infrastructure for studying over time genomic and environmental risk factors for heart disease, a disease that disproportionately affected African Americans.[85] Researchers would enter participants' demographic, epidemiologic, and plasma quantitative trait data into a "registry" along with their *ApoB* genetic data.[86]

No one involved thought the project would be easy. Since the days of the Tuskegee syphilis study, very little human subjects research had been conducted at Tuskegee, and the DARC gene research alerted all to the considerable ethical dilemmas. Everyone was concerned that the project be done right. One obvious challenge was deciding what could be done with the DNA samples. One reason to do genomic research in Tuskegee always had been that it was in the Black Belt, the region that Fatimah Jackson identified as best for collecting African American samples because the people there had deep African roots that could be traced through historical records. Since Jackson's original call, demand for these samples only had increased.[87] Extensive conversations took place between researchers at TU, UW, and NHGRI to determine whether samples collected by Tuskegee researchers might be used to meet these broader demands. While leaders in the genomics community at the national level were beginning to push for "blanket consent"—consent that would allow DNA samples to be used for many different research projects—no policy agreement yet had been reached.[88]

In addition to questions about sample use, the *ApoB* study raised questions about how to approach and recruit subjects. The DARC study highlighted the problem of offering money to subjects, as well as about the adequacy of informed consent. In the context of Tuskegee, it was crucial to ensure that subjects understood and freely agreed to the research. However, given the high poverty and low literacy rates prevalent in the Black Belt, achieving this presented significant difficulties.[89]

While challenging, the arrival on the TU campus of the National Center for Bioethics in Research and Health Care at the beginning of 1999 created optimism that these issues could be addressed. The center had been created from funds provided by the federal government as part of the apology for the Tuskegee syphilis study. One of the first research projects hosted by the center—the Communities of Color and Genetics Policy Project (CCGPP)—sought to include African Americans in the ethical and policy discussions about genomic research.[90] Within months of arriving, Ed Smith and the genetic counselor who ran the day-to-day operations of CCGPP, found common cause in their desire to find a way to work with the "grassroots community" of the Black Belt.[91] Talking only to members of community-based organizations, the genetic counselor feared, would bias her efforts, as members of these organizations came largely from the upper socioeconomic strata (SES). The problem for Smith was slightly different. He sought to reach those with low education and income levels because they were more likely to have lived in the region for a long time and possess the deep genealogical roots and stability needed for genomic research. While differing in their goals, their interests overlapped enough to write a grant proposal to create an infrastructure to educate the lower SES people in the Black Belt about genomics. The proposed infrastructure included a radio program broadcast on the

The old John A. Andrew Hospital closed in 1987 (photo by author).

Sign in front of the John A. Andrew Hospital announcing redevelopment for the National Center for Bioethics in Research and Health Care, April 2000 (photo by author).

local gospel radio stations that would make information about genomics available to those who could not read. Counties in the Black Belt also would receive a video about genomics research along with a video recorder machine and a television to play it on.

The Story in Tuskegee: Lack of a Hospital, Not Genomics

While this proposal put forward an innovative way of reaching people living in the Black Belt, it did not address the harder problem: would genomics be of interest to them? When I first arrived in Tuskegee, Alabama, in late April 2000 to learn more about the CCGPP, it would not be long before I heard the story about the Episcopal Church that caved in one sunny Saturday morning:

> That happened this summer, actually, just one Saturday afternoon, a day like this [a sunny day]. . . . It just caved in for no reason. And, I know we're getting off the subject, but . . . what if this happened on a Sunday afternoon when people were in church? I mean, it would be a major emergency, obviously. And we don't have any emergency care centers here in Tuskegee. So we'd have to take people either to Auburn or Montgomery, and the closest emergency facility has got to be about thirty miles. It's actually in Opelika, which is right past the other side of Auburn. . . . It's a commute between thirty-five to forty miles to get to a hospital. So that was a wakeup call, at least for me when I first moved here and heard about that. It's like, what if that happened a day later, twenty-four hours later, when church was on?[92]

I heard this story many times during my days in Tuskegee. It was a story about the lack of adequate health care, and it dominated discussion in the focus groups CCGPP was conducting. As one of the organizers explained: "Genetics, I think, can be seen as a luxury, being able to tell if your child's going to have Downs Syndrome, if you have sickle cell trait. Who cares if . . . someone has sickle cell trait as long as you have health care? The more pressing concern is health care, basic health care. Like the story with the church down the street, we don't have emergency room health care. . . . So even though most of our—all of our conversations—started out with genetics, they moved directly to basic health care."[93] Another put it more bluntly: "Let me say this, and I think this is real important, and it's the truth as I see it: it [genetics] is not important. And it is not important because there is *so much else wrong*."[94] In a town like

Tuskegee, where the poverty rate was over 50 percent, many had no interest in genetics.[95] "Nobody's got time to worry about this, or even think about it. It's a bunch of, you know, intellectual gobbledy-gook. It has no relevance to my life. I got to eat. I don't have a job. I don't have a house. My job is tenuous. I live from paycheck to paycheck. I have no assets. I've got to educate my children. I mean, black American life is about poverty, or on the edge of poverty. . . . I really don't think that anybody cares [about genetics]."[96]

Even the community leaders took persuading. Tuskegee simply had other needs:

> We've been drawn into begging these people to come [to the CCGPP focus groups]. And this is the privileged. We only have been working with one group of lower-income people. These are the upper-income people: the Episcopal Church, the social club, one of the top social clubs in the city. . . . And they've had to be made to understand why it should be important to them. But it's not important to them. I mean, there are other things. This town has no health care. In all this sophistication and stuff, we have no twenty-four-hour health care facility. We have no hospital. It is a half hour to any hospital. Or any acute care facility. So you are well-dead by the time you get there.[97]

Not only were people not interested, they were suspicious of genetics research—even of *talking* about genetic research: "Everywhere in the African American community people are very suspicious about anything with respect to research: Why are you doing it? What's it about? How's it going to benefit me? Or what are you trying to do to me now? You know, [the PI of the CCGPP project] starts this application with the question of people in Michigan when they ask, 'Just what are those people on the hill getting ready to do to us now?' "[98] The problem, as many in Tuskegee explained it to me during the time I spent there in April and May of 2000, was this: African American people are not against research; however, they are against research that might harm them, and they wanted to know what the possible harms and benefits might be, something the supposed experts had not done a good job of explaining.[99] Recalling a meeting held at the 1997 American Association for the Advancement of Science on the Human Genome Project titled "The Human Genome Project: What's the Public Got to Do with It?," a leader of the bioethics initiative at Tuskegee recalled: "It was an all-day discussion about what was happening, what the differences were between the HGP and the Diversity Project, and how to involve the community, or if to involve the community. And it was a fascinating thing, because nobody agreed on anything, and it was clear by the end of the day that nobody *knew anything*."[100]

Although the experts might not know what genomics might mean for the public, African Americans, she argued, did know some things: "What we do know, and what the average eighth grader can understand, is that based upon historical context this information looks as if it could lead to further stigmatization. . . . That is our history. . . . We know what happened with sickle cell, and the misuse of that information, and all the things that we had to do to try to reverse people's understanding. . . . [T]he trait is not a disease, and the trait is not a reason to discriminate against people."[101] These concerns were not unique to Tuskegee. When in the late 1990s Howard University formed a National Human Genome Center, researchers found that "fear of what's going to happen to my information, the genetics and DNA," to be a main concern in the community.[102] At North Carolina A&T in Greensboro, scientists told me of similar concerns. Members of their communities frequently asked them: What will happen to my DNA?[103]

The importance of these concerns could not be missed at Tuskegee. It was, as one African American scientist explained to me, "the scene of the crime." It was also now the site of a new national center devoted to repairing those past wrongs.[104] Problems could not be ignored. Worries and tensions on the ground when I visited were palpable. Although I had not come to Tuskegee to talk to people about the genomics research (indeed, I did not know about the DARC or *ApoB* research before I arrived) people were eager to talk to me about this research. The last thing someone said to me as I was leaving the Tuskegee campus was: "I know you are a good person. I know you will tell this story."[105]

A Story of Science and Justice

What was the story, and how could I ever tell it? For years, I struggled with this question.

Only two months after I visited Tuskegee, President Clinton proclaimed the "incandescent truth" revealed by the human genome: "all of us are created equal"; we all share a "common humanity." Francis Collins, leader of the public effort to sequence the human genome, and Craig Venter, leader of the private effort, echoed this sentiment in their remarks.[106] For these scientific and government leaders, the Human Genome Project had achieved the higher calling that the G-RAP researchers had envisioned a decade earlier: it had demonstrated our common humanity.

The story was moving and compelling. Indeed, how could one tell any other story?

Perhaps because if this story is ever to be true we must.

In 1978, partly in response to the problems made evident by the Tuskegee syphilis study, the National Commission for the Protection of Human Subjects of Biomedical and Behavioral Research published the Belmont Report. In this report the commission put forward three fundamental ethical principles that should underlie all human subjects research: respect for persons, beneficence, and justice. Over the following decades, bioethicists and lawyers labored over working out the meaning of the first two principles, but the last, justice, received little attention. As Georgetown University law professor Patricia King, a member of the commission, recalled in an interview in 2004: "We just didn't—we weren't able to work through some of these justice issues. . . . [W]e were terrific at understanding autonomy, informed consent, and pretty good at dealing with questions of beneficence. We just didn't have too much of a clue about what to do with this principle of justice."[107] For many, establishing informed consent as federal policy represented a major post-Tuskegee victory. However, it left unaddressed the more complicated questions the syphilis study raised. Although rarely discussed, the study presented Tuskegee doctors, scientists, and administrators with a difficult dilemma: participating in the research promised to bring resources to the struggling Tuskegee Institute and to the people of Macon County, many of whom did receive some sort of care because of the Public Health Service study.[108] It also promised the possibility of creating new knowledge about a disease affecting their communities. The dilemma then, as it is still today, is as King describes: "We are still struggling with . . . how to include minorities in [the benefits of] research without fostering or promoting stigma or biological understandings of race or conceptualizations of race."[109] While the US Department of Health and Human Services did create guidelines for how to include children, women, and prisoners in research, to date it has not produced a report on research on people who are educationally or economically disadvantaged. So, twenty-five years after the end of the syphilis study, when Tuskegee once again faced a decision about whether to participate in a major federally funded science project, fundamental questions remained unaddressed. Yes, Tuskegee researchers and administrators understood and firmly believed that informed consent would be required. Indeed, the Tuskegee Institutional Review Board (IRB) ultimately ruled that each use of a blood sample collected required informed consent. This decision went against then-emerging ideas about blanket consent. However, the question of informed consent was not the only challenge that genomics research at Tuskegee presented. More profoundly, this research raised questions about the proper practices and ends of science and justice in the twenty-first century.

It was clear to both the then Tuskegee president Benjamin F. Payton and the geneticist Ed Smith that the US government and corporations planned to invest in genomics as a fundamental new frontier in science, society, and the economy. However, how Tuskegee could adequately respond to its past while joining this genomic future remained unclear.[110] How could it uphold principles of informed consent, ensure that research met the needs of people, and take part in this exciting new domain of science and technology?

This was not just a question for Tuskegee. Tuskegee was never just about Tuskegee. As we will see in the chapters that follow, in the decade following the completion of the Human Genome Project, questions about how a diverse range of human beings could take part in human genomics in a meaningful and equitable manner became pressing for efforts to collect and study human genomes all around the world.

Should funding for genomic research, the benefits of which are unlikely to accrue to those with no or limited health care, be tied to requirements for funding basic health care? Who benefits from genomics research today, and on what basis can others be promised benefits in the future?

As the euphoria of the White House celebration of the human genome sequence wore off, these questions about the lived reality and meaning of genomics moved to the fore. Not addressing them would not make them go away. They only could be temporarily left behind as people and projects moved on to different places. In the fall of 2000, Ed Smith moved north to a better-resourced institution, Virginia Tech University. He took with him the sequencing machine and workstation that had been the beginnings of the genome center at Tuskegee. Plans for the Framingham-like heart study for African Americans—of which the *ApoB* study had been the start—moved from the historically black university of Tuskegee to Ole Miss, Mississippi's predominantly white flagship university.[111] And a federally funded effort to map human genetic variation moved its focus from the race-riddled terrain of the United States to the hoped-for humanistic frontier of the globe.

4 WHO REPRESENTS THE HUMAN GENOME? WHAT IS THE HUMAN GENOME?

When people have to walk their donkeys fifteen kilometers one way to get water, and you are looking at ground that is red-orange and you pick it up and it runs through your hands like sand, you look at what should be a green garden and there is nothing there, and the cattle are dying . . . what does it mean for me to put out my arm and give you blood?
—Interview with International HapMap Project Community Engagement leader

African Americans asked to participate in genomic research would not be the only persons concerned about genomics' liberal policies of inclusion that failed to address questions of justice. Many Native Americans that NHGRI consulted raised concerns that the authority of genomic research might be invoked to threaten tribal sovereignty as well as divert attention from more pressing health care needs.[1] While NHGRI originally hoped that recruiting human subjects in the United States would steer their efforts clear of charges of biocolonialism sparked by the Human Genome Diversity Project, these concerns raised by some African American and Native American leaders made it evident that collecting in the United States would be no panacea. This along with funding from other countries led leaders at NHGRI to once again consider the possibility of an international initiative.

This move also aligned with broader political shifts. The opening years of the new millennium marked a high-water mark of hope for a new global order that could transcend old divisions. The Cold War had ended. Brazil, Russia, India, and China (BRIC) were emerging as new powerful actors on the global stage. The Internet was making global communication a daily reality for many. The human genome promised to reveal our common humanity not just on the national level within nations but also internationally, between them. In the midst of this turn-of-the-millennium optimism about the new worlds opened up by global connection, many leaders in the field of human genomics felt strongly that the effort to map human genetic variation should be an international effort.[2]

Yet launching a worldwide effort to map human genetic variation—what became known as the International HapMap Project (HapMap)—posed significant problems. In addition to worries about a repeat of the Human Genome Diversity Project debacle, there were significant practical challenges. The globe is a vast place. Not everyone's genomes could be surveyed. As one program officer at NHGRI put it to me: "The ultimate resource would be to sample everyone in the world. You can't do that."[3] So who in the world should be sampled?

While this question had no clear answer, NHGRI leadership did have a clear idea about who should not be sampled: "small, isolated" indigenous populations. The HGDP, they believed, ran into trouble not because of inherent problems surrounding studying the differences between human genomes but because HGDP leadership decided to sample small "vulnerable" populations.[4] Success, they hoped, might come through sampling what they described as "large populations" who held majority status in society. This approach promised to be both scientifically and ethically feasible. At the scientific level, recent research revealed that the structure of the human genome contained blocks knowns as haplotypes in which genetic variants strongly associate with one another. Genetic researchers believed that these structural blocks were similar in diverse populations, and so sampling only a few "major" populations from geographically diverse locales would be enough to create a map of human genetic variation of use for biomedical research.[5] Ethically, sampling a few major populations also promised to reduce the chances of discrimination. Majority populations, so the theory went, were more likely to enjoy the powers of self-governance and education that—within a liberal worldview—secured the ability to represent and govern oneself. To help ensure the ethical robustness of this approach, leaders of the HapMap decided to extend to these populations new rights to participate in crucial decisions about the naming and sharing of their DNA samples.[6] This decision followed a well-worn path of American liberal democratic governance. The United States had long sought to live up to its promises of democratic freedom for all people through forging new paths for including those historically excluded from the polity.[7] Initially, many embraced the HapMap leadership's efforts to extend these liberal practices of inclusion to genomics through including communities in decisions that had formerly been the domain of scientific experts. Yet quickly challenges arose.

In line with liberal assumptions, leaders of the HapMap presumed the existence of people organized into "communities" who could in effect be delegated some limited powers to represent and govern the samples collected.[8] Yet, as we will see, in an age of global transformations this presumption faced many challenges. The end of the Cold War, free trade agreements, and the rise of digital communication all fostered novel relations that crossed the traditional bounds of government, reconfiguring who people were and who could legitimately represent them. Genomics grew out of these transformations, contributing to them powerful new practices for differentiating among human beings that generated novel questions about who the "people" might be who could take up the task of representation.

The absence of any clearly demarcated people to govern linked to an unraveling of a consensus about the proper constitution of their rights. While HapMap organizers positioned informed consent as the most important of these, many of the "communities" they sought to sample raised more basic questions about what it meant to take part in the creation of a global map of human genetic variation and what entitlements and rights should come along with participation. Echoing the views of the citizens of Macon County, communities from Ibadan, Nigeria, to Houston, Texas, believed that that their willingness to take part entitled them to resources—such as medical clinics—and access to the latest biomedical research. Bioethicists, NIH leaders, and researchers struggled to respond. The stories of their efforts bring to life the more fundamental questions of justice and knowledge raised by this first major initiative of the postgenomic era.[9]

Who in the World Can Represent the Human Genome?

In the summer of 2001, NHGRI began efforts to create a map of human genetic variation, what NHGRI leader Francis Collins called "the single most important resource for understanding human disease, after the sequence."[10] While the United States would be the first to commit funds—initially $40 million—discussion soon ensued about whether the United States should take the lead. As one bioethicist involved in the planning recalled: "There was a heated, albeit brief discussion, about doing it just in the United States. . . . Lord knows we have a gracious plenty of diversity here *and* we have a set of rules for how we do research."[11] Administrators at NIH worried as well that working with other countries on a project that required international sharing of human tissue samples and data might prove challenging, and in some cases impossible.[12] China and India, for example, as part of their effort to defend their national sovereignty in the realm of genetics, prohibited shipping the DNA of their citizens outside their national borders. Human rights issues in China and Japan posed additional difficulties.[13] All of this led at least one NIH program officer to prefer a US-based effort: "My own preference would have been to have done all the sampling in the US precisely because administratively it would have been a whole lot easier, avoiding all these problems associated not just with the human rights issues in China and Japan, but also just the complexities of international collaborations. It's really, really hard to do this."[14] However, planning for the HapMap began right at the height of belief in the power and importance of the human genome and global cooperation. To do the project only in the United States struck many as exclusionary. One cultural anthropologist who worked on the HapMap later reflected:

> It would have been very easy [to do it in the United States]. In fact, you can get brown people, black people, white people, and all different versions of our genetic heritage. You don't have to leave [a major US city] to find that. But that doesn't make me happy . . . because I want to see a bigger playing field. . . . If you start to gatekeep in that way, then it becomes one more time when the rich countries get to have the science, they get to have the benefits of the science production, they get to develop the technologies, they get to produce the drugs.[15]

Some argued more forcefully. Reportedly, genome scientists in Japan argued that if the United States took over the project of creating the next generation of genome maps it would amount to nothing short of imperialism.[16]

Despite deep concerns created by the Diversity Project, it was clear that the times had changed, and for "the project to be real it had to be international."[17] And so early in the planning of the initiative, NHGRI leadership decided to include researchers in Japan, China, Canada, and the United Kingdom.

Yet in the eyes of many, the United States and the NHGRI clearly remained at the helm, providing the largest amount of funding as well as coordination for the project. The Diversity Project also still loomed large in the memories of many. Thus from the very start the HapMap leadership realized that it faced a worrying question: How could a global survey of human genetic variation primarily funded by the United States avoid charges of racism and colonialism?

In consultation with many of the scientists and ethicists who had been involved in the Diversity Project debates, NHGRI forged a two-part strategy. First, collect samples from "large" genetically variable populations. These populations, organizers reasoned, did not face the same risks of discrimination as those faced by the so-called small vulnerable indigenous populations that the HGDP sought to sample.[18] Second, conduct community engagements prior to sampling DNA. That would give communities the opportunity to discuss issues and raise concerns as well as exercise new powers to decide how their DNA samples would be labeled and made available for research.[19] Following this two-pronged approach, HapMap organizers hoped they would not "fall into the Diversity Project trap."[20]

They were partially successful. They did avoid the challenges of researching and working with indigenous populations. They did not avoid the broader, more fundamental questions about how to select and define human groups for sampling in a just, democratic manner. Partly, this is because they did not anticipate the degree to which the nature of these questions had changed since the days of the Diversity Project. In the intervening years, leaders of human genomics successfully promoted a vision of human genomics as a twenty-first-century global science that would not divide, but unite. To the surprise of many involved with the HapMap, rather than a risky endeavor that should be avoided, many sought representation in human genomics. From Denver, Colorado, to New Rochelle, New York, to the Tuscan region of Italy, "communities" the HapMap approached reported feeling honored to be chosen to take part in the HapMap, an international science project of such great import.[21] Problems arose not as NHGRI sought to include these groups, but as they made decisions to not represent them in the next generation of human genome maps.

Genomics had become a high-prestige, high-stakes science to which many sought entrance. Representation in the next generation of human genome maps did not present risks so much as it promised membership in twenty-first-century global societies and knowledge economies. Thus, how genomics leaders should identify and define the people who in practice should be represented, as well as define the rights that should accompany representation, became significant points of contestation. These issues moved beyond American bioethics' liberal approaches that sought to include research subjects in an ethical manner to more fundamental *bioconstitutional* issues of how to *constitute* citizens and their rights in twenty-first-century biosocieties.[22]

Constituting Communities as Representatives of DNA

Learning from the experiences of the Human Genome Diversity Project, leaders at NIH did anticipate that efforts to study human genomes would require new processes for defining subjects and their rights and set out to formulate policies to implement them. Judith Greenberg, a program director at the National Institute of General Medical Sciences (NIGMS), took the lead, convening in September 2000 "The First Consultation on the Responsible Collection and Use of Samples for Genetic Research." Since 1972, NIGMS had maintained a collection of cell lines at the Coriell Institute. Prior to 2000, almost all of these cell lines derived from the cells of people affected by disease and "the apparently healthy."[23] However, with the completion of the Human Genome Project and the growing desire to study genetic variation, NIGMS received increased requests for cell lines not from *diseased individuals*, but from *diverse populations*. While the demand was great, NIH leaders expressed caution. The Diversity Project controversies demonstrated that collecting from populations raised special concerns. The September 2000 consultation that I attended confirmed for NIGMS that these concerns were widely held. As the executive summary of the consultation explained: "The potential for discrimination, stigmatization, and breaches of privacy is a major concern of individuals and communities who may wish to participate in the research. Other concerns relate to the definition of communities, the perceived benefits and risks of the research for these communities and their members, and the full participation of communities in the entire research process. Different communities and community members will have different needs and interests in the proposed research and may want to participate in various ways."[24] While these concerns were not universal—notably, representatives of disease groups argued against additional oversight and for wider use of samples—leaders at NIGMS believed that on the whole the message was clear. As a program officer later explained to me:

> INT: We had our marching orders . . .
> JR: Marching orders from the meeting?
> INT: Yes.
> JR: From the communities saying we want to be consulted.
> INT: Exactly.[25]

From the viewpoint of the leadership of the NIGMS cell repository, communities had spoken: they wanted a voice in decisions about how their DNA could be collected and used.

Yet not all agreed with them. As one NIGMS leader later recalled, some scientists "really objected strenuously" to giving communities a say in how samples were collected and used. They asked: "Why should any population have anything to say about these samples? After all, we are the scientists and we know."[26] However, others felt strongly that if the human genome was a powerful molecular phenomenon that could reveal important information about individuals, then those individuals should have rights to govern it.

Advisory meetings at which the community consultation policy was discussed proved contentious. It quickly became clear that there was a difficult line to walk between those scientists who wanted to control the samples and those who believed communities must have some control. Reportedly, the leadership of a high-ranking African American NIH official helped moderate the tension and enabled the creation of NIH guidelines for best practices.[27] These became the basis for the "Policy for the Responsible Collection, Storage, and Research Use of Samples from Identified Populations for the NIGMS Human Genetic Cell Repository."[28]

This policy instructed those who sought samples to "consult with members and leaders of the community" and "assure that those consulted *are representative of the socially identifiable population* from whom samples are to be obtained" (italics added). It further stipulated that consultations "should involve interviews, focus groups, discussions, town meetings or other forums." They should clearly present the goals of the research, describe who could use the samples and the length of time samples will be stored, describe risks and benefits, and explain that the community could withdraw samples at any time. In addition to the consultation, the NIGMS policy called for the creation of a "community advisory group" that would provide ongoing governance of the samples. Anticipating the need for long-term governance, it stipulated that the community advisory group should "decide upon the term of an individuals' membership and develop a plan for replacing members." It required the director of the NIGMS cell repository to "maintain contact with the community advisory group for as long as the samples remain in the repository" and to consult with the group "about any proposed use of the samples that in any way appears to differ from that originally agreed upon." Finally, it mandated that the results of research should be "disseminated to the community in an understandable way."

On the question of whether communities would have power to shape what research was done using their samples, however, the policy was less than clear. It stated only that community consultations should "provide information on whether the community will be able to specify certain types of research for which samples may not be used." It also provided incomplete guidance on the troublesome question of how samples would be named, only advising researchers to avoid naming populations unless populations' names were "scientifically significant" or "the community (through the community advisory group) wants to be identified."

Despite this vagueness, for the time, the policy was bold and far-reaching. It laid out specific responsibilities for researchers and rights for communities and created an ongoing governance structure for the samples. However, for NHGRI and the leaders of the HapMap, the policy did not go far enough. Because NHGRI planned to deposit the cell lines created for the HapMap into the NIGMS Human Genetic Cell repository at Coriell, at the very minimum it needed to abide by the NIGMS policy.[29] However, rather than just meet its minimum standards, administrators at NHGRI led an effort to clarify and strengthen it.

Specifically, NHGRI specified that communities would have the power to provide "direct input into the informed consent process and into decisions about how the samples collected from their community will be collected, described, and used."[30] Importantly, a document distributed at a HapMap planning meeting in 2002 stated that "communities themselves will ultimately decide the exact description or label by which they wish samples from their population to be identified."[31] For some at NHGRI, changing the name from community consultation to community engagement marked these more substantive powers offered to communities.[32] As one innovator of the policy later reflected: "Engagement conveys what we thought we were actually doing: going out and truly having an equal partnership. . . . [C]onsultation still has this very separateness to the way that the relationship could be structured."[33]

For many, the HapMap's community engagement policy introduced a significant new approach to human genomics research. Community engagement took seriously the problems raised by the Diversity Project and introduced concrete practices designed to ensure that participants in NHGRI's efforts to create the next generation of human genome maps were not treated as objects of study, but rather as subjects with the rights and power to shape the study itself.

The power to self-identify became most emblematic of this shift.[34] It, along with the right to control the terms on which the samples remained at Coriell, convinced some who once had been critical of the Diversity Project to sign up to help organize the initiative. As one leader of a community engagement later reflected: "The community advisory groups having that much power made some administrators very uncomfortable, but I thought that was the coolest thing to be participating in: to absolutely inform communities what they're getting into and then having the power over their own samples."[35] A leader of one of the NHGRI's community engagements in Tokyo, Japan, also remembered the early enthusiasm for the HapMap's community engagement policy: "There was [*sic*] high hopes internationally that the community engagement and ELSI approaches for HapMap would be the best ever used in international genetics research."[36]

Who Is the Community? The Limits of Liberalism and the Return of Race

Yet while pushing efforts to democratize genomic research into brave new waters, community engagement did not overcome the limits of American liberal approaches to governing that shaped the efforts at both NIGMS and NHGRI. As committed as both institutes might be to devolving responsibility for governance from scientists back to the people, for the effort to work there needed to be a mechanism to identify and represent the people—what NIGMS and NHGRI called "the community"—who could take up their powers of self-governance. In the United States, and among administrators whose careers unfolded in the post–civil rights era, it is understandable that NIH leadership assumed these mechanisms existed or could be built. During the 1960s, in the midst of the civil rights movement, inclusion of formerly excluded groups emerged as a dominant policy paradigm.[37] Organizations such as the National Association for the Advancement of Colored People (NAACP) facilitated these changes through taking up the work of representing African Americans and other minority groups in the negotiation and implementation of laws and policies, most notably the Civil Rights Act of 1964. NIGMS leaders had this US context in mind when they crafted their community consultation policy. Recalled one NIGMS administrator: "The way we wrote the guidelines for the repository we were really only thinking about the US populations."[38] In the end, though, these were exactly the groups that the HapMap did not want to sample, favoring sampling large "majority populations" abroad rather than "minority populations" in the United States.

Community engagement practices carried forward this mark of the American context to Japan, Nigeria, and China, creating challenges. Majority populations did not need to organize themselves into clearly defined "communities" in order to gain access to resources and representation. While one could debate whether there exists an "African American" community in the United States, and whether organizations such as the NAACP could represent it, there simply was no equivalent "Japanese community" in Japan about which to even begin a debate. The Japanese in Japan formed the nation; they were not a community. Even in places where the HapMap did seek to sample members of "large populations" who had immigrated to the United States, it was hard to find and define communities. One could try and hang out and meet local people at churches and shops, but as one cultural anthropologist recruited to do this work later observed: "You can only go grocery shopping so much. It doesn't make sense after a while."[39]

Thus, while HapMap leaders offered new powers to "communities," it proved far from self-evident who these "communities" were who could take up these new rights. If these communities were to exist, HapMap researchers, ethicists, and those asked to give their voices and DNA to the HapMap would be part of constituting them, a process that would involve novel biosocial acts. As one community engagement leader later reflected: "I began the engagement there saying . . . I'm here to come to your picnics, go to your masses and funerals, and to talk to you about whether or not you would donate blood for a study."[40] These were strange new formations indeed. How and on what grounds they should proceed proved far from clear.

These complexities are absent in the deceptively simple description of the initiative that appears on the official HapMap website:

> The International HapMap Project is analyzing DNA from populations with African, Asian, and European ancestry.[41]

Although a seemingly clear directive, it proved highly contentious. The social science and ethicist observers of the initiative quickly noted that "African, Asian, and European" strongly evoked racial systems of classifications. As one NIH program officer observed: "I think that there was an awareness from day one that this would be seen by many people as essentially a race map."[42] This proved a major challenge. Whatever the grounds were going to be on which HapMap populations and communities would form, they could not be the grounds of race. The NHRGI knew all too well that the Diversity Project failed to move forward because many believed it would reinvigorate racial science.[43]

HapMap leaders sought to avoid similar problems through distinguishing between what they described as "largely social constructs such as race" and the population names they used.[44] They argued that the former arose from subjective social practices that were prone to bias, while the latter derived from the precise methods of "sound study design." As the HapMap official website explains:

> The way that a population is named in studies of genetic variation, such as the HapMap, has important ramifications scientifically, culturally, and ethically. From a scientific standpoint, *precision in describing the population* from which the samples were collected is an essential component of sound study design. . . . From a cultural standpoint, precision in labeling reflects acknowledgement of and respect for the local norms of the communities that have agreed to participate in the research. From an ethical standpoint, precision is part of the obligation of researchers to participants, and helps to ensure that the research findings are neither under-generalized nor over-generalized inappropriately. (Italics added)[45]

HapMap leaders envisioned an integration of science, culture, and ethics into a precise approach for naming populations. Importantly, they did not believe that they would play any role in constructing populations. In their view, only races were socially constructed. Populations were biologically real; they only needed scientists to precisely describe them.

In interviews with me, and in their own writing, several genome scientists stressed this commitment to precision and credited it to social scientists. Before, they loosely referred to "Africans" and "Europeans" and "Asians." Now they used descriptions guided by the goal of "precision" that drew on objective criteria, such as the scientists' sampling criteria. Exemplifying this approach, in one interview a HapMap researcher explained to me that in the case of the sampling done in Africa, the project is "not studying Africa" or even "all the Yoruba; it's the ninety DNA samples from the Yoruba that we had."[46]

However, such a position posed a significant problem: to claim the object of analysis is only "the ninety DNA samples from the Yoruba that we had" was to insist on a degree of precision that would render the HapMap useless. As another project organizer explained to me (this time with reference to the sampling done in China):

> INT: If you sampled Han in Beijing Normal [University], and you sampled in Shanghai, and you sampled someplace way out west, and they were just terribly different from each other, and the tag SNPs [single nucleotide polymorphisms] from one just had no bearing on the other, then unfortunately we would have done a HapMap that had these four little points that were sort of useless for any other population.
>
> JR: Right. So in that sense, the samples collected in Beijing weren't just representative of those people, but . . . they were representative of something bigger than that, or the HapMap wouldn't have worked.
>
> INT: That's right.[47]

As this organizer implies, the governments of the United States, Japan, China, Canada, and the United Kingdom did not spend tens of millions of dollars to go to China or to Nigeria to learn something about forty-five or ninety people (the number of people sampled in each place respectively). These people were picked with the hope that they would represent a large group of human beings. However, because the question of representation in human genetics (and now genomics) was historically vexed, understandably many organizers of the HapMap sought to avoid the issue altogether by either appealing to precision to define the populations sampled, or by declaring that "the Project [did] not aim to define populations."[48]

What's in a Name?

While project leaders rejected the idea that they sought to define populations, they did accept that in practical terms they needed to label the samples collected from them. For this task, they turned to "communities." Recall that granting HapMap communities the power to name their samples led many to believe in the democratic potential of the HapMap. Yet in practice this devolution of power from scientists to people proved anything from straightforward. This became evident as soon as the collections started. Citing the case of samples collected in Japan, a program officer at NHGRI explained: "There were a lot of people who were saying they wanted the samples to be called 'Asian.' Which is fine, except—Well, there are a couple of issues. One, you also have samples from China, so are they 'not-Asian'? Or they're just 'China'? You have 'Asian,' and then you have 'Chinese.' . . . So this notion of let the community decide for themselves what to call themselves is not quite so simple. Everybody assumes they're going [to] come up with a label that's scientifically defensible, which it really isn't, because Japanese people obviously don't represent all Asians."[49] These problems led to a change in the HapMap leadership's approach to naming the samples. Rather than give communities the "ultimate" voice in deciding the names, as stated at a planning meeting in the fall of 2002, the official paper on the project's ethical approach published in *Nature Reviews Genetics* in 2004 explained that communities would "have *some input* into how they wanted their population named" (italics added).[50]

Yet even with these reduced powers, it was not clear that researchers were prepared to work with HapMap designated communities. As one social scientist who oversaw a HapMap community engagement explained to me:

> INT: The community was supposed to have input on the names—the labels. . . . So the first time I came back, [my community engagement coordinator] told me . . ."You know, they came up with these most amazingly creative names."
>
> JR: Like what?

INT: Traditionally red . . . which was red for blood, red for wine, and red for left wing. . . . I mean . . . we had a list of about seven or eight fascinating names. HapMap leaders in Washington, DC, however, rejected all of these names.

I asked why.

INT: Too idiosyncratic, nobody would know what they were. . . . They wanted some relationship with Europe.[51]

In interactions with project leaders in Washington, DC, it became clear that HapMap "communities" would not have "the freedom" to choose their own names according to their own criteria. Instead, they were bound by a notion of precision that was meaningful to genome scientists: precise geographic location. So they tried again, this time suggesting names that met the scientists' desire to have a reference to Europe:

INT: They came up with Europa en [this region of Europe]. I said to them, "Europa en [x]?"—You know, it's very interesting. My logic of the people in [name of the town] would be that this would be X en Europa—

JR: Yeah.

INT: They said, "No! Europa en X." They said, "after all, they've been invading us for years!"

None of the names were satisfactory, and the only thing the community in the end chose was that their name be printed in their native language in addition to English.

That the community did not have "the freedom" to choose their name did not upset this engagement organizer. Indeed, she admitted to being "very jaded about this free choice crap." In any case, "it was an experimental project and things change all the time."[52] What bothered her was that leaders at NHGRI insisted that HapMap "communities" would have a free choice to choose their names even after it became clear in practice that they would not:

INT: I just had so much trouble getting them to own up to the fact that they were going to put parameters. I couldn't get them to acknowledge that.

JR: Did they just keep saying "they can choose"?

> INT: Yeah. "They can choose, but this isn't okay, and that isn't okay." And they wouldn't frame it like "within these parameters they can choose." They wouldn't own up to the fact that they were constructing the field, or that they were going to have a say in the field.[53]

Despite the original desire to grant an autonomous space, free from the genome scientists, in which HapMap communities could determine their names, no such space existed.

Social and political theorists have spent much time explicating the problems created by theories of the right of autonomy that do not take into account the social and political conditions in which these rights are imagined and granted. Perhaps most illuminating for the purposes of understanding the challenges that HapMap organizers faced is feminist political theorist Judith Butler's discussion of recognition and naming of the self in *An Account of Oneself*: "A regime of truth offers the terms that make self-recognition possible. . . . This does not mean that a given regime of truth sets an invariable framework for recognition; it means only that it is in relation to this framework that recognition takes place or the norms that govern recognition are challenged and transformed."[54] As HapMap communities learned, their views and beliefs would receive recognition only insofar as they were legible within genome scientists' regime of truth. This regime, as then NHGRI director Francis Collins later reflected, involved computers and their standards of nomenclature: "I think the scientists wanted to have a name so they can go into the computer and pull them out and know what they are, which means that they cannot be three paragraphs with ten footnotes. It's got to be short and content rich. But I don't think the scientists had a particular bias about what the answer should be within those constraints about it being machine usable. Exactly how that plays out, that ought to be the community decision. It must not be chosen, though, in a way that's confusing."[55] HapMap communities could not simply freely decide their names. Neither could the HapMap simply grant new rights to communities. Neither rights, nor names, nor communities, nor populations existed in advance for the HapMap to use, study, or respect. Instead, the HapMap leaders inevitably played an important role in constituting them.

These were important acts. They played central roles in deciding who would be recognized as making up the world in the new powerful science of genomics. The HapMap "community" in Japan sought to represent "Asia." Another in southern Europe sought to make visible the history of Western invasions. HapMap leadership wanted to avoid "confusion," and so embraced "precision," which they defined as using specific geographic locales to represent those they sampled: Tokyo, Japan; Ibadan, Nigeria; and so on. HapMap leadership understood that these were important decisions and devoted considerable resources to addressing them. However, the tools of American bioethics—tools that assumed the prior constitution of people and their rights—proved poor instruments for navigating these constitutional issues. While they provided an ethical framework, they failed to address a deeper question of justice: who and what matters in today's biosocieties? Nothing would make this more evident than the HapMap's effort to construct a robust process of informed consent.

From an Ethic of Informed Consent to Justice in a World "In-Formation"

When proposers of the Human Genome Diversity Project met in State College, Pennsylvania, in October 1992 to decide which indigenous populations they hoped to sample, no representatives of these groups attended the meeting. They had not been invited. Indigenous rights organizations subsequently cited this meeting as evidence that the project viewed indigenous peoples as mere objects of research and not subjects who needed to be informed and involved in the research process.[56] Thus, not surprisingly, when NHGRI leadership set out to create its own initiative they put great energy into creating a robust informed consent process.

Key to their efforts was an attempt to break down the divide between researchers and research subjects. Influenced by the cultural anthropologists they consulted, they acknowledged that research might present different risks and benefits to different communities, but the only way to find out would be to ask communities. Thus HapMap leadership decided "to give people in the localities where donors were recruited the opportunity to have input into the informed consent and sample collection processes."[57] They implemented focus groups in which they asked HapMap designated communities about their concerns. They wrote informed consent forms that represented these concerns.

Despite a change made to facilitate mutual respect, my interviews with community engagement leaders revealed a level of uneasiness about the process. On the ground, those who led the focus groups did not find people eager to talk about their views on the human genome. For many, the genome simply was not a matter of concern. Indeed, it did not exist. "I would come to the United States and I would hear 'Genome, genome, genome, genome.' And I would go back to [country X] and they would say, 'Genome what? Genome who? Genome whatever.'"[58] In this community, the genome had no lived reality, nor was it clear how it could or should. This problem of the reality of the genome also arose in a resource-poor area of Africa where the HapMap approached a community to collect samples, but found that the members of the community did not want to participate. "So what's real? I mean, when people have to walk their donkeys fifteen kilometers one way to get water, and you are looking at ground that is red-orange and you pick it up and it runs through your hands like sand, you look at what should be a green garden and there is nothing there, and the cattle are dying . . . what does it mean for me to put out my arm and give you blood? So issues of access and relevance definitely did come up."[59] While NHGRI leaders recognized that different people might have different views about the risks and benefits of learning about the human genome, they did not anticipate that people might have different understandings of the very reality of the human genome. In order to get people to the table sharing their views about something, it must first be a thing—that is, a matter of concern.[60] In 2002, when the community engagements started, for many people around the world, the genome was not something about which they had concerns for the simple reason that it had no lived reality for them. Thus, not surprisingly, community engagement organizers often faced difficulty getting anyone to show up to the focus groups and town hall meetings they organized. In one locale, despite putting advertisements up all over the subway system, almost no one came. In another, people had more important engagements to go to:

> INT: The last meeting we had took place on a Saturday, and there was a demonstration against some environmental problem. Tons of people went there.
>
> JR: That's too bad.
>
> INT: Well, that's where their priorities are. That's what I am saying.[61]

When people did show up, they often showed little interest in their new right to inform the informed consent process and asked the leaders of the community engagements to explain to them what issues mattered. As a cultural anthropologist who led one of these engagements explained: "In my prior work there was no question of translating. I was asking people questions about things that they were already intimately familiar with. Here, on the other hand, I'm asking them about things that, one, they probably never thought about. A lot of the times they had not, so much so that they would return the question to me: 'So, you're asking me a bunch of questions. You tell me something about it. I don't know squat.'"[62] Yet often even the academics with PhD's who led the community engagements could not answer the questions. Even basic issues such as the question of how the HapMap was producing a map proved hard to answer.[63] This lack of understanding did not come from failing to possess the relevant expertise. Instead, genomics was in formation. How genomic variation should be understood and represented—as single nucleotide variation, as blocks of variation, as a map—was a live matter of debate, even among genome scientists. Given this, the process of informed consent met many challenges.

Indeed, in many cases it seemed out of place. The right of informed consent rose in importance in post-Tuskegee United States. In the wake of the revelations of deception in this US-government-funded syphilis study, and the suffering that the men went through because they participated in this research study without being informed of or asked to consent to it, understandably many felt informed consent was of the utmost importance. However, since the 1970s, the meaning of biomedical research has changed significantly. Rather than a potential source of harm that one sought to avoid, by the late 1990s, as the efforts to create a genome sequencing center in Tuskegee made clear, many scientists and policy makers viewed inclusion in biomedical research as an important right that assured people inclusion in the latest biomedical research and treatments.[64] In the case of genomics, inclusion in human genomics also promised the right to "participate in defining what it means to be human."[65] In several of the places that NHGRI sought to collect samples for the HapMap, the right to be represented in the new biomedical sciences mattered more than the right of informed consent. Indeed, one leader of a community engagement reported: "Nine times out of ten people dismissed [informed consent] as legal mumbo jumbo."[66]

While many of the people who took part in community engagements had little interest in the informed consent process—and the discussion of the risks and benefits of human genome research it entailed—other elements of the HapMap initiative did interest them: the HapMap's status as an international science project; leadership of the initiative by a world superpower, the United States; blood donation. These things did have meaning for people and guided the discussion away from informed consent and toward more fundamental questions about citizenship, nationhood, and participation in twenty-first-century global societies.

For many, the act of giving blood invoked these deeper ties of belonging. Deepa Reddy, a cultural anthropologist who led the community engagement for the collection of the HapMap's "Gujarati Indians in Houston, Texas" samples, explained the historical and political roots of this tie between blood and citizenship in the formation of the Indian nation: "Indira Gandhi's oddly premonitory statements about living and dying just before her assassination in November 1984 and her specific allusion to blood—'I can say that each drop of my blood will keep India alive, will make India strong'—then become material to encourage blood donation in her name and in the name of the cause (India) she represented in those final moments of her life."[67] Giving blood for the HapMap, Reddy explained, provided "a route to citizenship," a way to participate in "a project of surely Gujarati, surely Indian, quite possibly South Asian, and then potentially global significance."[68] Members of the Indian American community in Houston, Texas, understood that the US government considered the creation of human genome maps to be an important project of the United States. They sought full recognition as citizens of that nation. So just as they had fought for a statue of Gandhi to be placed in Houston's most important public park, so too did they advocate for the inclusion of their DNA in the new lands being chartered by the human genome.

Informed consent proved a poor instrument for recognizing these bonds to larger political collectivities for its forms ultimately asked people to act as autonomous individuals. Many resisted this liberal imaginary. As one organizer of a community engagement later recalled: "I had one woman say to me, for example, why should anyone think that my sample had anything to do with me? Why should I think that my blood has anything to do with me?"[69] Reportedly, many gave blood for the HapMap for the good of higher forces: the nation, the globe, God. These acts of service, of citizenship, and of sacrifice brought great meaning to their lives. Thus, significant disappointment resulted when those originally approached to be HapMap communities—in Denver, Colorado, in Houston, Texas, and in New Rochelle, New York—found out that their samples would not be collected or used for the official International Haplotype Map.[70] NHGRI approached these communities before its leaders ultimately decided to collect outside the United States. After the decision to create an international initiative, collecting in these communities became a much lower priority, and in some cases, never took place. As a leader of the community engagements in a Catholic community on the East Coast explained to me, members of the community were still talking about their disappointment when she visited sometime after the HapMap had come to an official close: "There is this Christian saint who says that what we are about in this world is gathering up drops of the savior's blood, meaning gathering suffering. And they [members of the community] said [I] was actually going to gather [their] blood. So they still remembered it, and I think there was some disappointment that they weren't involved."[71] For many of the people asked to take part in the HapMap, the request to give blood was an honor: an invitation to the community of nations, the community of God, the global community. For some, recognition by the US government in and of itself possessed great value: "One of the focus groups was all clients of the social service agencies—so people who had undergone electroshock therapy, who had no teeth, who barely had more than one set of clothes. They came in incredibly angry at the government, distrustful, and left in tears hugging me saying: 'No one's ever listened to us before. It's really phenomenal.'"[72] For many, taking part in the HapMap meant much more than furthering research; it meant recognition as a respected member of the polity, and representation in twenty-first-century global biosocieties.

Blood for a Hospital?

Yet what rights and resources should accompany representation in twenty-first-century biosocieties remained far from clear and became a source of much discussion and deliberation among HapMap leaders. As traditional natural resources—such as land and oil—dried up, and states at the end of the twentieth century started to look to their citizens' blood and tissues to create biological resources, the makeup of biological bodies became a new ground on which citizens began to make claims on the state.[73] These claims moved beyond the bounds of bioethics and onto the terrain of biological citizenship. Thus, perhaps it should not have surprised HapMap organizers when informed consent—even a robust informed consent process that engaged communities—proved inadequate for mediating the relationship between US scientists and those in resource-poor countries from whom they sought blood and DNA. And yet reportedly it did come as a surprise when one such HapMap community asked the NIH to support a hospital: "We were at a meeting with the advisory council, and one of the advisory board members read out loud an unbelievable letter essentially saying, 'We're honored that we've been selected to represent our people in this project and as sign of reciprocity, we want a hospital.'"[74] By the time of this meeting, members of the community had signed informed consent forms saying that they understood that they would not benefit from participating in the HapMap. They also reportedly communicated to NHGRI leadership that they were honored to represent their people in the HapMap. Thus, they explained, they did not anticipate a request for a hospital.

However, the request did fit squarely within what by that time had become a recognized norm for biomedical research: in resource-poor areas research should address local health needs. Specifically, in 2002 the Council for International Organizations of Medical Sciences (CIOMS) published *The International Ethical Guidelines for Biomedical Research Involving Human Subjects*. Guideline 10 of this document states: "Before undertaking research in a population or community with limited resources, the sponsor and the investigator must make every effort to ensure that: the research is responsive to the health needs and the priorities of the population or community in which it is to be carried out." It further elaborates: "It is not sufficient simply to determine that a disease is prevalent in the population and that new or further research is needed: the ethical requirement of 'responsiveness' can be fulfilled only if successful interventions or other kinds of health benefit are made available to the population."[75] No one would dispute that the HapMap would lead to no direct health benefits for HapMap communities, while a hospital would.

While the request fit within international guidelines, it was hard for the United States' National Institutes of Health to respond. As an NIH program director reflected: "NIH rules just don't contemplate that. . . . So its like on the one hand we say we want to be responsive to community concerns—and I think in this case we were—but it wasn't without a huge amount of just having to fight every step of the way in the NIH bureaucracy. And it's not that there's anyone there who's trying to be evil, it's just that there's these rules that are really hard to get around."[76] A branch of the US federal government, the NIH was prepared to recognize individuals and to fairly compensate them for the labor of going to the clinic and giving blood. However, its rules did not recognize social collectivities and inequalities among them. Eventually, after extensive machinations and going "way up in the NIH infrastructure," an exception was made and a modest amount of funding money provided to expand the hours of the existing clinic. However, as a program officer close to the process observed, the case was not exceptional: "As long as there's gross disparities in resources, this issue is going to remain problematic and controversial."[77]

Indeed, the issue did arise more than once over the course of the HapMap, creating great uneasiness. NIH researchers and staff wanted to help. They sought to respect the wishes of communities, yet they found that the bureaucracies in which they worked constrained them.

Bioconstitutional Challenges

While the HapMap emerged at a time of great optimism about the power of globalization and genomics to create a more equal and just world, in practice the institutional methods and theories needed to realize this vision proved hard to forge. As one lawyer who played a central role in the HapMap later told me, no "good political theory" existed to guide the endeavor.[78]

Certainly, no theory existed to decide who research subjects represented. While HapMap leaders initially assumed their research subjects would be members of already-formed communities and populations, the effort to name their samples brought out widely diverging ideas about how the human species could be carved up into representable groups. Human collectivities simply did not exist out there for the HapMap to name and study. They did not come already constituted, ready to participate in the governance of the initiative. This made the democratization of genomics more complex than just including "people" in research design and regulation, as much of the democracy and science literature at the time recommended.[79] To include people, first organizers of the HapMap would need to take part in constituting them.[80]

The liberal democratic framework that informed American bioethics and the HapMap's approach to ethics failed to anticipate this problem.

In its focus on the ethical inclusion of people in research, this framework presumed the prior existence of people. Thus it did not address the bioconstitutional questions that faced the HapMap: Who are the people who should have the right to be included and represented in maps of the human genome? What should that right of representation entail? HapMap leaders presumed—as did the organizers of the Human Genome Diversity Project—that their social and cultural anthropology colleagues who studied the makeup of societies could answer these questions. Yet as these colleagues reported back their findings from the community engagements, at times they faced significant challenges from genome scientists. At issue were questions about how human beings should be defined and cared for in the genomic era, and who had the expertise and rights to answer these fundamental questions.

At issue as well were questions about the constitution of the human genome. The HapMap proceeded from the assumption that the human genome was a powerful molecular structure that played an important role in human disease. It emerged during an era of strong support for the Common Disease–Common Variant hypothesis (CD-CV), the hypothesis that common genetic variants linked to common disease. It was on the basis of hypotheses such as the CD-CV that HapMap leaders sought to extend new rights to research subjects: if one's genome contains powerful and potentially stigmatizing information about oneself, then one should have the right to govern access to it. Yet as the decade unfolded, challenges to belief in the power of the human genome and the validity of the CD-CV hypothesis became more frequent, leading some to question the legitimacy of creating exceptional new rights for research subjects. As one bioethicist who helped to organize a HapMap community engagement argued: "It really strikes me as genetic exceptionalism at its core to say that this stuff is so dangerous that we have to have a special community review. I think there's a lot more stuff out there that is much more damaging than that."[81] The HapMap's bioethical practices did not anticipate, nor could they respond to these questions about what a genome is, and what powers it possessed. Instead, they presumed that the human genome was a powerful molecular phenomenon. It was—and should be—a matter of concern around which people gather to understand, debate, and govern.

Some people approached to donate samples did not concur. They barely had heard of the human genome. It certainly was not a matter of concern. What did concern them were the inequities that existed between the US National Institutes of Health and their local health infrastructures. NIH invested billions of dollars into studying the human genome, a novel scientific object of uncertain medical relevance and meaning, while the health infrastructures that could make a meaningful difference in their lives remained underfunded.

The HapMap was at the leading edge of a decade that brought to the fore these questions of justice. These questions moved beyond the everyday workings of the NIH bureaucracy. Health care bureaucracies are not in the business of forging fundamental new rights and forms of governance. Like all forms of bureaucracy, they are supposed to implement the will of the governed, not constitute it. The HapMap's novel governance structures, while founded on millennium hopes for novel forms of global governance, ultimately lacked a political theory and source of authority. Historically, nations, not biomedical research institutions, possess this authority. Thus, perhaps it should not surprise that alongside the International HapMap several national initiatives arose to make sense of the human genome.

5 GENOMICS FOR THE PEOPLE OR THE RISE OF THE MACHINES?

Nothing comes from nothing, nothing ever could.
—Rodgers and Hammerstein, *The Sound of Music*

Nations offer built-in cultural and political practices for managing questions about how to value and sort human differences, something experienced daily by millions who attempt to cross their borders. Beginning in the mid-1990s, genome scientists began to turn to these national infrastructures to help them answer questions about how to define and govern populations for the purposes of studying their genomes. Genomes, in turn, appeared to promise nations the opportunity to create new natural resources out of the bodily tissues of their citizens at a time when resources in their lands and seas were disappearing.

In the late 1990s, several genomics initiatives emerged that sought to take advantage of the confluence of these political-economic and scientific goals: deCODE in Iceland; the Estonian Genome Project; the UK Biobank.[1] Because of the emphasis it placed on the principle of public sovereignty, notable among these initiatives was Generation Scotland (GS), a genetic/genomic study of "the people of Scotland."[2] In 1998, the people of Scotland overwhelmingly voted to establish an independent Parliament. The new government in Edinburgh opened its doors in 1999. That same year Scotland's chief medical officer and chief scientist endorsed the initial plan for GS.[3] Early GS documents aligned the initiative with the devolution movement. Collecting and studying the genomes of the Scots, they argued, would bolster public sovereignty, a cornerstone of modern liberal democracies.[4]

Yet while promising that GS would shore up, not challenge, the principle of public sovereignty that motivated the efforts to devolve power from London to Edinburgh, in practice studying the DNA of the people of Scotland raised novel questions about the makeup of a sovereign people and their rights of self-government. In Scotland as in other nations that sought to study "their own" DNA, the value of nationally sourced genomes ultimately required scientists from across the globe to use and access them.[5] Would sending Scottish DNA across national boundaries threaten a key component of Scottish sovereignty, control of its natural resources?[6] This question arose in the midst of broader debates about what ties to London should remain, and what movements across the border of Scotland and England should be allowed in a post-devolution Scotland.[7]

At the urging of the new government, GS initially decided that the genomes collected in Scotland would not leave the country and would only be studied by Scottish researchers. This decision was consistent with the goal of building an initiative that embodied the moral sentiments and values of the people of Scotland. Unlike a similar effort in Iceland that raised the ire of critics who charged the Icelandic government with selling the genomes of its people off to an international pharmaceutical company, GS leadership sought to act in a manner that was in line with the interests of the Scottish people. This did not mean that GS would deny pharmaceutical companies access to the GS resource. However, it did mean that GS would engage the Scottish public in decisions about monetization and ensure that all the biological samples it collected remained under direct physical control of the nation.

Yet how to maintain these commitments in the midst of the shift from genomics to postgenomics proved elusive. In the years that followed the completion of the public-funded Human Genome Project, private funding took on a much greater role in human genomics. Sequencing technologies became a significant site of venture capital investment, sparking the innovation of next-generation sequencing technologies. While Edinburgh had long been a hub of human genetics research, keeping up with these "next-generation" machines proved beyond the reach of publicly funded research initiatives, especially after the worldwide financial crisis hit in 2008. Thus the leaders of Generation Scotland faced a dilemma. Taking full scientific advantage of the DNA samples they had collected from Scottish citizens required analyzing them on the latest machines—machines that now sat outside Scottish borders in hubs of technological development such as Oxford, Cambridge, and Silicon Valley.

Had genomes proved as valuable as first thought when the initiative launched, the pressure to resolve this dilemma might not have been so great. However, by March 2011, when GS collected its last DNA sample, the scientific, medical, and economic worth of Scottish DNA was not clear. Thus the central question was no longer how to fairly adjudicate access and benefits. Instead, it was how to transform Scottish DNA into a thing of value. At a minimum, more sequencing than originally imagined would be required. Access to the latest sequencing machines became vital.

Efforts to resolve this issue brought into relief the challenges of knowing and governing in the networked, transnational, and entrepreneurial worlds of postgenomics.[8] These worlds are global in reach but still must respect and negotiate national boundaries. They give people a voice in how to fairly distribute the value of genomics but struggle to create this value. Efforts to address these postgenomic challenges brought GS into unchartered waters in which they no longer could presume the worth and meaning of either the genomes they had collected or the democratic practices they had created. By the time the initiative began to collect its first samples in 2006, the heady days of the HGP and the early years of the HapMap were over. The value of both human genomes and novel democratic approaches to governing them now were uncertain. In the wake of these shifts, GS faced a difficult question: Would it support Scottishness, or was Scottishness itself "in-formation" along with genomes and the sequencing machines they required? A fundamental reevaluation of GS' approach to ethics and knowledge followed. Whether liberal democratic, humanist, or cosmopolitan values should govern the initiative hung in the balance. How to make genomes valuable became a practical concern.

I followed along as these debates unfolded. I first arrived in Scotland in September 2005 to give a plenary talk at the launch of the United Kingdom's Genomics Policy and Research Forum.[9] Over the course of the decade that followed I made several return trips, spending the winter and parts of the summer of 2008 based at the Forum in Edinburgh, as well as shorter trips in 2010, 2012, 2013, and 2014. Over that period, I interviewed many of the main leaders of GS, most more than once, and followed the evolution of Scottish devolution, ending with the dramatic Independence Referendum vote in 2014.[10] In what follows, I tell the story of Generation Scotland in the context of these changes, paying particular attention to the new questions of knowledge and just rule sparked by the postgenomic problem of value and meaning. Before arriving at these questions, however, we first need to understand how it came to be that questions about the value of Scottish genomes became caught up in questions about the values and meaning of a new Scottish government.

National Genomes?

Why would a nation attempt to transform its people's blood and DNA into a resource of value? While in times of war and in other moments of national need, it is not unprecedented for citizens to offer up their blood, attempts to use biology for political ends have a vexed past.[11] Eugenics and sterilization policies figure as dark moments in the histories of many polities.[12] Yet in the opening years of the twenty-first century, some nations began to consider the possibility that far from oppressing its people, collecting DNA and tissues from them might promote their sovereignty. In Estonia, for example, genomics promised a way to transition from command-and-control Soviet-style economies to free-market knowledge economies associated with western European democratic states. Acting on this hope, in 2000 the Estonian Parliament passed legislation that established the Estonian Genome Project.[13] In September 2004, in an analogous effort to join the global knowledge economy, Mexico launched the National Institute of Genomic Medicine.[14] Perhaps most famously, in December 1998 the Icelandic Parliament passed the Act on Health Sector Database which permitted deCODE Genetics Inc. to link the health and genealogical records of Iceland's citizens with their genomic data, a key step in transforming Icelandic genomes into a new natural resource.[15]

The effort to transform genomes into a national resource in Iceland is particularly telling because of the hopes raised and controversies stirred.[16] For Iceland, genomics was high stakes. The home of the world's first national parliament (formed in the year 930), over the centuries Icelandic people fought to defend their sovereignty and to defend the small island nation from takeovers by Denmark, Norway, England, and Germany. They also struggled to establish a diverse and stable economic base. While historically fishing provided the majority of the nation's economic base, fish in the sea were being depleted. deCODE Genetics and its founder, neuroscientist turned Harvard-trained genome scientist Kári Stefánsson, successfully convinced a majority of those in the Icelandic parliament, the Althing, that genomics could help address these long-standing challenges. Genomes offered the nation a new natural resource that could make Iceland a major player in the then-emerging international bioeconomy.[17]

Stefánsson's promises soon began to bear fruit. In February 1998, Hoffman-La Roche agreed to pay deCODE up to $200 million to receive the rights to work with the company to develop small molecule drugs.[18] deCODE and Stefánsson convinced La Roche that like the populations that organizers of the Human Genome Diversity Project had proposed studying, the Icelandic people were relatively isolated and homogenous; thus, they constituted an ideal population for finding genetic variants associated with disease.[19] Additionally, they were "educated" about the value of science. Thus, some scientists argued, they did not present the same ethical challenges raised by the Diversity Project.[20] As a prominent population geneticist in Europe explained to me in 1999: "You have much less of a problem with informed consent if you are dealing on the whole with an educated population."[21] Or so it would seem.

As it turned out, what constituted being educated and informed about deCODE proved a major point of controversy in Iceland and worldwide. With the passage of the Act on Health Sector Database, the Althing ruled that its citizens would have the right to opt out of, not into, sharing their medical records with deCODE. Stefánsson defended the decision, arguing that it would not be practical to get all 270,000 Icelanders to consent to be a part of the database.[22] Leading biomedical researchers and the Icelandic Medical Association, with the backing of the World Medical Association, opposed the decision. They argued that turning the medical, genetic, and genealogical records of the Icelandic people into a commercial asset raised grave concerns.[23] University of Washington population geneticist Mary-Claire King concurred, dubbing deCODE a "colonial treasure hunter."[24]

These concerns about deCODE's commodification of the human genome reached a broad international audience. In January 1999, the *New York Times* published a commentary written by Harvard population geneticist Richard Lewontin in which Lewontin claimed deCODE would transform the "entire population [of Iceland] into a captive biomedical commodity," hardly the act of a democratic society.[25] Perhaps most pointedly, Stanford law professor Henry T. Greely asserted: "This isn't fish or lamb's wool we are dealing with here. This particular product is the spirit of the human species, and before Iceland turns its DNA into a commodity. . . . I just hope they know what they are doing."[26] Iceland might presume to own and control the salmon and trout in its seas, but DNA represented a different kettle of fish.

deCODE launched during the high-water mark of the genome era, when many believed that genomes were objects of great value. The controversy it sparked foreshadowed the crisis of value that would soon come to characterize the postgenomic condition. While many agreed that DNA was a thing of great value, what kind of thing it was, and what kind of value it possessed, proved uncertain. deCODE with the approval of the Althing, sought to treat DNA like another natural resource. Scientists, doctors, and ethicists both within and outside Iceland objected. DNA, they argued, inhered in us all, revealing our connections to one another. It was nothing short of the "spirit of the human species," not amenable to control by a company or a nation.

The debate in Iceland took place at a time when both scientific and political leaders invoked the human genome as an agent of change in the transition from a world carved up into discrete nation-states that struggled to control finite resources to one world guided by cosmopolitan ethics and shared concerns.[27] As US president Bill Clinton pronounced at the White House celebration of the completion of the draft of the human genome: "I am happy that today the only race we are talking about is the human race."[28] This was a powerful ethical vision. Yet on the ground, how to enact it remained far from clear. What kind of a thing was the human genome, and who had the right to know and control it? A nation? A corporation? All humans?

Generation Scotland: A Social Contract for Genomics?

The experiences in Iceland played a formative role in shaping how these questions would be answered in Scotland. Like in Iceland, researchers in Scotland argued that the Scottish people constituted an ideal population for medically relevant genomic research.[29] Further, like the Althing, the new Scottish government sensed an economic opportunity. Genomes promised a new natural resource to replace oil in the North Sea. They also promised a source of revenue to support the new Scottish nation. However, unlike in Iceland, Scottish researchers and government officials could not avoid the broader questions about the ethical valence of collecting blood and DNA from citizens. The deCODE controversy made headlines in the United Kingdom.[30] Additionally, in the late

1990s several high-profile public scandals rocked the British biotechnology and biomedical sectors, including the widely publicized protests against genetically modified organisms.[31] Most relevant to Generation Scotland, in 1999 the Alder Hey organ scandal broke. The British public learned that a pathologist working at Alder Hey Children's Hospital in Liverpool removed without permission human tissues, including the organs of children.[32] In the wake of these scandals, British institutions implemented policy changes designed to "engage" the public in decisions about the governance of science and technology.[33]

These developments resulted in a large public investment in "science and society" initiatives in the United Kingdom right at the moment that scientists and policy makers began to seek support for Generation Scotland.[34] Indicative of these investments, the first grant proposal the GS team submitted asked for support for a robust public engagement plan. Indeed, this is what originally sparked my interest in Generation Scotland. Not only did GS plan an extensive effort to discuss the initiative with various segments of Scottish society, and solicit their views on how it should be governed, they also promised to listen to and respond to the findings. As the GS website stated: "GS has made a promise to listen, discuss, and respond to the findings; not just around ideas about study materials, recruitment or participation but also on the more difficult social, legal and ethical issues."[35] This promise positioned GS as different from other efforts. Social science researchers in the United Kingdom critiqued both deCODE and the UK Biobank for either not attempting to understand public views and interests or for doing so in a superficial manner.[36] As a result, they argued, deCODE faced public opposition and the UK Biobank struggled with low response rates. GS, by contrast, set out to be different. As members of GS explained to me:

GS 1: We're opt in. They were opt out.
JR: Yes, and is that significant for Generation Scotland?
GS 2: I think it is.
GS 1: It was key.
GS 2: Yes.[37]

In Scotland, unlike in Iceland, the people would have the power to decide whether to opt in to providing their data and DNA for genomic research.

Beyond this, GS leaders committed to creating a project that expressed the interests of the Scottish people. Through extensive focus groups, surveys, and interviews, they proposed to understand and respond to the views of the Scottish public.[38] Tapping into Scotland's formative role in the Enlightenment, they believed empirical research could reveal these views. The results of this research could then be used to guide the creation of a social contract for GS that represented the values and beliefs of the Scottish people.[39]

This notion that Generation Scotland would be a project of, for, and by the people of Scotland is apparent from the very opening screen of the GS website. In 2008, the GS homepage greeted its viewers by flashing, in order, the following three phrases:

> The Health of the Nation
> A Unique Partnership
> Our Genes, Our Health, Our Future[40]

Another section of the website announced: "Generation Scotland is a partnership with all the people of Scotland."

The Scottish context offered a number of ways to make the case. First, Scotland long had been known as the "sick man of Europe." Government studies consistently found national rates of cancer and heart disease to be the highest in Europe.[41] Given these public health challenges, it was possible to argue that a genomic study focused on these diseases would benefit the Scottish people.[42] As one GS organizer explained to me: "If you want buy-in for a population study of health, if those health problems are prominent in that population then you are going to get more buy in from the population and from the government. And if you look at the words that were used in the Genetics in Health Care Initiative . . . [it] talked about the big three: cancer, mental health, and stroke."[43] Indeed, in the early years of the project, GS leaders frequently cited this commitment to serving the needs of the people of Scotland. As Professor Andrew Morris, then chair of the Scientific Committee for Generation Scotland, explained in a press release announcing the recruitment of the one-thousandth volunteer for the project: "We hope this study will help to unlock the secrets of Scots' health and bring real health benefits to those living with disease and to the next generation."[44]

Second, in addition to better public health, attention to public consultation also helped to distinguish GS as a Scottish project. Scotland has a long history of collectivist government.[45] GS sought to build on this history, and to ensure that the project embodied the will of the people.

Third, the legacy of the Scottish Enlightenment and concomitant commitments to reason helped to justify a new Scottish commitment to scientific and medical research.[46] As the "About Generation Scotland" page of the GS website began: "Scotland has a strong medical tradition that continues to the present day. Scotland trains proportionally more doctors and undertakes more biomedical research than the rest of the UK." Generation Scotland sought to take advantage of this and "the Scottish public['s] . . . very strong tradition of supporting medical research."[47]

Finally, the very notion of a Scottish people with a will distinct from that of the rest of the United Kingdom gathered strength in 1998 when the Scots voted to establish a separate Scottish Parliament. Emerging at the same moment, GS shaped and was shaped by this broader effort to establish a sovereign Scotland.

On the Health and Wealth of Nations

As part and parcel of this effort, the newly formed Scottish government set out to establish an independent Scottish economy that could support the nation, and Generation Scotland offered a new potential source of wealth at this pivotal moment. The 2002 Royal Society of Edinburgh Discussion Dinner about Generation Scotland begins with the claim that GS would link wealth and health.[48] The guidelines for the Genetics Healthcare Initiative that eventually funded GS also envisioned developing Scottish research that would "grow strong Scottish businesses" and "meet the needs of the people of Scotland."[49]

At the time, joining health with wealth appeared viable. On April 3, 2006, Scotland's health minister announced a collaboration with what was then one of the world's largest pharmaceutical companies, Wyeth Pharmaceutical Co.[50] Wyeth agreed to invest £33 million over five years to develop a Translational Medicine Research Collaboration with an option to extend for a further five years. Reportedly, Generation Scotland played a major role in Wyeth's decision to invest in Scotland. For many pharmaceutical companies at the time, genomics offered the possibility of targeting drugs for use in particular populations.

This "targeted" approach to drug development might have been of particular interest to Wyeth. Only a few years earlier, the NIH's Women's Health Initiative reported that Wyeth's hormone replacement therapy (HRT) Prempro significantly increased the risk of invasive breast cancer.[51] While the data against HRTs looked insurmountable, others gave Wyeth and Prempro a "second chance." If the "right drugs are used with the right regimen in the right patients," a 2003 *EMBO Reports* article argued, HRTs could "indeed be beneficial to women's health."[52] GS's goal of genotyping individuals from a wide subsection of a population aligned with the hope of finding those "right patients."

Yet while the Translational Medicine Research Collaboration made sense for Wyeth, it would only work for Scotland if it appeared in accord with the will of the Scottish people. As one member of the Scottish government explained to me: "We had to make sure that whatever model [of access to the GS resource] we eventually move to is one that the public are happy with."[53] This, the government understood, was not just a matter of the initiative's medical and economic value. As the case of deCODE made eminently clear, deriving the wealth of a nation from its genomes was also a matter of public morality.[54]

The importance of the link between morality and economics would not be lost in Scotland, home and birthplace of Adam Smith, author of the *Wealth of Nations*. Although it is largely forgotten by the many who remember Smith for his metaphor of the "invisible hand" and his contributions to "free market" philosophies, Smith was a moral philosopher who believed that the advantages of commercial society only could be enjoyed if society upheld the principles of equity and justice.[55] Smith did not view money and wealth as sins but rather as practical tools of nations that could improve the well-being of its citizens if nations respected these ethical tenets.[56] However, Smith also read with interest and sympathy the words of his contemporary, Jean-Jacques Rousseau: "In a genuinely free state, the citizens do everything with their own hands and nothing by means of money."[57] For Rousseau, money did not help citizens; it threatened the bonds between them, and thus the social contract.

Two centuries later this tension over the moral valence of money and wealth played out in the debate over whether a truly free society could make money off of its citizens' DNA. Natural and social scientists, lawyers, and policy makers in Scotland did not escape these tensions, but in the tradition of Smith and the other philosophers of the Scottish Enlightenment, they pursued a practical resolution grounded in empiricism.

Ethical Empiricism: Discovering the Will of the Scots

Like Smith, the leaders of GS did not see money and the pursuit of profits and wealth as necessarily a corrupting force. As long as the moral views and values of the society were respected and upheld, "moral economic solutions" could be found.[58] As a member of the Scottish Executive explained to me, the challenge then was to understand these values: "Rather than disguise it, you best to say to people 'This is what we are aiming to do. What are your views about this? How can we do this so it's most acceptable to you?' . . . So for example, if you've got the resource and Pharma come, should you be saying, 'We will do that work for you and give you these results?' Or do you let Pharma directly mine the resource?"[59] For answer to these questions, GS leadership turned to social scientists.

Distinguishing Generation Scotland from the deCODE case, sociologists of science and medicine took part in GS from the very start of the initiative, and held seats on the first GS governance body, the Scientific Committee.[60] Their primary role was to empirically investigate the beliefs of the Scottish public about GS and the issues it raised. These findings would inform GS policies.[61] For the social scientists involved in the project, this latter point proved essential: unlike consultations done for UK Biobank, which many critiqued for following a "market research model," the views of the Scottish people would directly shape the policies adopted by GS.[62]

An empiricist model acceptable to both the natural and social scientists facilitated this approach.[63] This model presumed that social scientists could produce data about Scottish people's moral sentiments and that lawyers then could create laws to practically implement them.[64]

But how could social scientists know the moral sentiments of the Scottish people, and the Scots' views on Generation Scotland? For this, GS social scientists employed a mixed-method approach: focus groups, in-depth interviews, public surveys, exit questionnaires, and ethnography. Each method, they reasoned, compensated for and built upon the limitations and strengths of the other methods. Focus groups revealed the range of issues relevant to the diverse segments of Scottish society but could not go in depth or aspire to be representative. In-depth interviews probed issues in more detail, but still were not representative. Surveys expanded the number of people you could reach and represent, but could not go in depth.

Based on the focus group research, GS social scientists reported that while most Scots recognized a need for pharmaceutical companies to develop drugs, the majority believed that access to Generation Scotland's DNA database should be restricted to medical professionals and academic researchers. Second, most felt it important that the database be publicly owned. Finally, they believed profits should be shared so that they benefited the community.[65] A public survey of 1,001 members of the "general public" created to determine the Scottish public's "preferences for Generation Scotland study design" confirmed and refined these results.[66] According to this survey, the issue that made the most difference to people's willingness to take part in GS was whether benefits would be given back to the National Health Services (NHS) or charities.[67] This replication of results in both the focus groups and the survey, as well as the strength of the preference expressed in the survey, led the social scientists to conclude that while Scottish people were willing to have pharmaceutical companies involved in GS, they wanted profits to be shared with the public through some form of benefit-sharing agreement.

To put this finding into practice, GS required a mechanism for benefit sharing. To create this mechanism, the social scientists came together with the lawyers to create a sociological justification, as well as a legal framework, for benefit sharing.[68]

To this point, all had worked as envisioned: the views of the Scottish people had been ascertained, and a practical mechanism for expressing their views had been constructed. Yet the question still remained: would the GS leadership respond and implement benefit sharing and create a project that reflected the values of the Scottish people?

Implementing the Will of the Scots

From the start, social scientists involved in GS believed strongly that their work could not just be window dressing.[69] It needed to lead to concrete policy changes. Benefit sharing proved a test case. Yet, as a case, it proved inconclusive. In late 2008, during my second GS fieldwork trip, concerns arose about whether GS would support benefit sharing. In the eyes of some GS leaders, a data access policy had been posted to the GS website that contained no benefit sharing policy. Some asked for the data access document to be taken down until a benefit sharing mechanism had been worked out. For practical reasons (the need to apply for Phase II funding) others argued to keep the data access policy up.

Yet what the data access policy did contain at this time hints at a deeper issue: the meaning of benefit sharing. In Appendix 6 Section 3.1. of the GS Management, Access, and Policy Publication (MAPP), a section on "Income Sharing" states: "Following deductions (e.g., for patent and legal costs) the net revenue derived from commercial exploitation shall be divided between the parties as follows: 25 percent to the University parties; 25 percent to the NHS parties; 25 percent to the party/parties which own the commercialized project rights; 25 percent to support the Generation Scotland initiative."[70] For some of the GS leadership, sharing profits and funneling them back to GS and the NHS did represent benefit sharing and demonstrated how the public consultation had shaped GS.[71] One organizer of GS later referred to this approach as "true" benefit sharing: "It [benefit sharing] is referring to the fact that some of the financial return is going back into research, so it's true benefit sharing."[72] However, for the lawyers and social scientists who conducted the community consultations and developed the benefit-sharing proposal, the issue was less clear. This version of benefit sharing, while a step forward, was very different from the one the lawyers and social scientists who worked on GS described in a 2007 *Social Science and Medicine* paper. In this paper, they proposed a model based on one developed in Newfoundland and Labrador in which researchers were required to do the following: "Submit a proposal for benefit sharing to the Standing Committee on Human Genetics Research, along with a rationale for the proposal, and final approval for the research project would only be granted if the Standing Committee were satisfied that an adequate benefit-sharing arrangement had been made in light of the following principles."[73] These principles included distributed justice, respect for the communal nature of the information contained in DNA, and promotion of health as a common public good. A simple carving-up of the profits fell short of determining benefit sharing in a manner that upheld these three principles. While they believed some progress had been made, some social scientists felt the GS policy was not enough: "I mean I was kind of disappointed at the time. And they did include benefit sharing in . . . the participant information leaflet . . . but it wasn't the kind of benefit sharing that was coming up in the focus groups or the kind of benefit sharing that [we] had written about."[74] GS's policy of funneling money back into research was part of the vision of benefit sharing, but it was only one part. More could be done. As one of the GS social science and legal team later reflected:

> I . . . think entities like Generation Scotland and UK Biobank and other biobanks around the world have real opportunities to engage much

> more imaginatively in issues of benefit sharing. Because if . . . this evidence which is coming out in GS and other studies indicates that the prospect of commercialization and profit payoffs are a concern, then it seems to me that there are things that can be done very specifically about sharing benefits . . . about making it clear and strategic about how this resource is actually benefitting broader communities. Because obviously it's easy to say, well we want the medical research to be done and that might give rise to medical products or medical advances. That's one form of benefit. But there may be other forms of benefits that could be shared if commercialization becomes a reality, though it's not clear whether commercialization is going to become a reality so we're still dealing in the realm of the hypothetical.[75]

An expanded notion of benefits presented in the *Social Science and Medicine* article included jobs and "free access to any test or treatment."[76] However, it is this last issue—whether GS could even be commercialized—that became the more pressing issue, and one that ultimately pushed benefit sharing into the background.

So far GS, like deCODE in Iceland, has generated no profits, and as one GS organizer explained to me in the summer of 2012: "For there to be benefit sharing there needs to be benefits and there currently are none."[77] The Scottish government originally funded Generation Scotland during the Human Genome Project era, amid great belief in the potential value of genomes. However, by the time GS had collected and processed the DNA and health data of participants (the last blood samples were processed in 2011), the value of genomes had dropped, and the economy had crashed, taking down with it Wyeth (which Pfizer bought in 2009). The major pharma investment in GS that the Scottish government had anticipated evaporated. Consequently, the main problem GS faced was no longer pharmaceutical access and what to do with profits. Instead, it was how to sustain the GS resource in the first place. Rather than distributing funds, it needed funds to survive. In light of these postgenomic developments, one of the social scientists later expressed understanding for why a richer form of benefit sharing had not developed: "I think there was a motivation there, but I think at the end of the day, the Generation Scotland project had to work, and they were trying to get the bloody thing going."[78] For that to happen, the GS resource had to itself be of value.

Under these conditions, the pressing question was no longer who might access and develop Generation Scotland's valuable genomes and associated data and tissues, but "How could Generation Scotland develop the samples and data it had collected to give them value?" Further, just as GS could no longer presume genomes of value, it also no longer could assume the existence of a distinct people, the Scots, who possessed clear values. This became evident as GS attempted to answer a critical question about how to develop the value of its resource: Could Scottish genomes leave Scotland?

Postgenomic Dilemmas: The Decline of Scottish Genomes and the Rise of Global Machines

As part of its original vision of how Generation Scotland would serve the nation and its people, early on the Scottish government advised, and GS pursued, a policy of keeping the biological tissues collected by GS in the country. As one GS leader put it, the government understood the biological tissues, in particular the DNA, to be "a national treasure" that they did not want others to "plunder."[79] Instead, the Scottish government hoped that developing and mining the DNA, data and urine collected by GS would help build the Scottish economy. As one GS organizer explained: "We are funded by the Scottish government . . . and by having sample analysis done within Scotland it is actually creating jobs in Scotland and it is supporting Scottish enterprises by keeping it here."[80] In the wake of the Iceland controversy, the decision to keep the samples in Scotland also fit with the commitment to public sovereignty—that is, to holding the samples in public trust, and not private interest. As one of the organizers explained to me in 2008: "If we were to have done that [send the samples out of the country], we would have had to have considered the confidentiality, data protection laws in that country, and the protection of those samples. We're trying to keep control of the GS resource and samples on behalf of the participants. We're the custodians of the samples and to do that, we want them to be kept in this country."[81] Finally, authors of the 2003 *Generation Scotland Legal and Ethical Aspects Report* suggested that framing the project as Scottish might motivate participation in GS.[82] Promising that biological samples would not leave the country and that only Scottish researchers would study them followed from this suggestion.[83] The *Participation Information Leaflet* given to potential subjects during the first phase of recruitment stated:

"Research using the samples will be done by staff in Hospitals and Universities throughout Scotland."[84]

When GS launched, dominant theories about how to find medically relevant genetic variants in the human genome appeared in harmony with this desire to keep samples in Scotland.[85] The Common Disease–Common Variant hypothesis, which predicted that most of the genetic risk for common, complex diseases arose from common variants, still held sway.[86] Thus project leaders believed that to find these variants GS would not need to sequence all three billion nucleotides of the genomes they collected, but only the common variants (less than 0.1 percent of these genomes). This task proved well within the capacity of the existing technology at the Genetics Core of the Wellcome Trust Clinical Research Facility in Edinburgh where GS stored and analyzed its DNA samples. Thus, keeping samples in Scotland for analysis appeared viable.

However, by the time GS began to actively advertise and make available its resources in 2011, belief in the CD-CV hypothesis had waned. Researchers had found few common variants that could be associated with disease, and many now believed that rare variants would be key to understanding disease. Finding these rare variants necessitated sequencing whole genomes, which meant the technological requirements of genomic breakthroughs increased exponentially.[87] Rather than sequencing one million points in the genome where common variants were located, one now needed to sequence all nucleotides, or all the nucleotides in the coding regions, increasing the amount of data needed by orders of magnitude.

This change—combined with the end of a single public-funded effort, the Human Genome Project—spawned a new private industry that sought to build low-cost, rapid, and accurate sequencing and analysis methods that could deal with these increased sequence data needs.[88] Driven by venture capital and a demand for innovation, these so-called next generation sequencing methods changed at a rapid rate, making it nearly impossible to keep the latest machines on site. Instead, keeping up meant moving your samples off site to the machines that could "run" them the fastest.

Historically, many of those machines were located in the United Kingdom. Fred Sanger (after whom the Sanger Institute is named) developed the Sanger sequencing method that a generation of biologists and geneticists used. Indeed, I used Sanger sequencing methods while

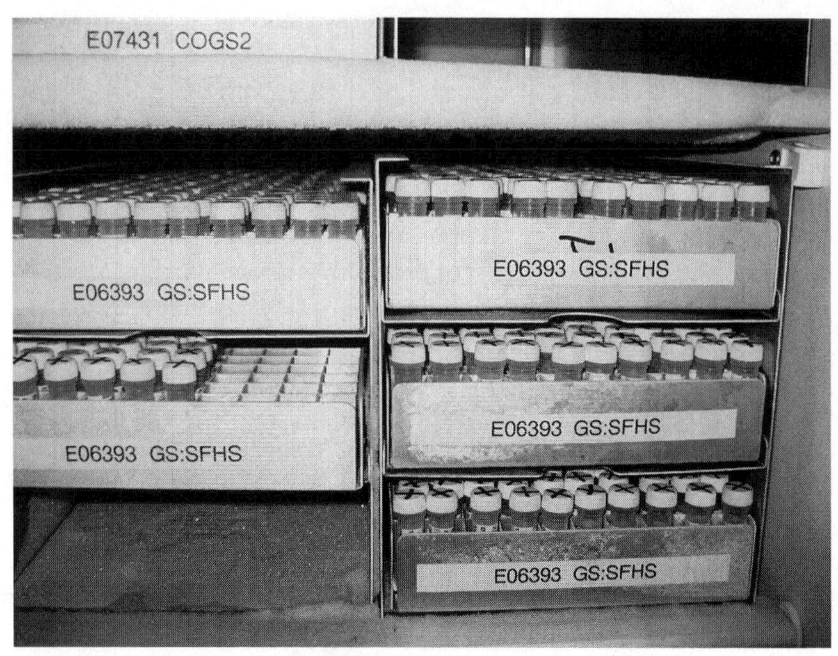

Frozen Scottish DNA samples (photo by author).

working in developmental and molecular evolution labs in the early to mid-1990s. However, in the years following the completion of the HGP, genomics went global in an intensive way. Silicon Valley, Beijing, Boston, and other places around the world began to compete with Cambridge, England, as the home of the fastest machines. The days when an international public project such as the HGP could set standards and stabilize technology platforms quickly became a thing of the past. A competitive, private, fast-paced and high-stakes game soon eclipsed this public approach. As one of GS organizer explained: "Ever since I have been in this line of work, the genomics technology has been ramping up. I think it is fair to say the last three, four years really there's been a steady change, and I would say up until that time, relatively modest core facilities could genuinely see what we're offering as the state of the art. If we want genotyping done, we can do it. I think those days are gone. I think with the high throughput sequencing, so-called next generation sequencing, you're either in that game or you're not, and we are not."[89] Thus it was not surprising that when I returned to Scotland to interview

GS leadership in 2012 after the recruitment phase was over, GS faced a dilemma. It had told participants it would not ship their DNA samples outside of the country, but that's where the fastest machines now were. What was to be done?

Who Are the Scots?

First, one had to figure out what was meant by the promise in the *Patient Information Leaflet* to not send the samples out of the country. For one GS organizer, it meant that they could not, for example, send their samples to China, and for some obvious ethical reasons:

> There's sort of factory level genomic analysis being done in China in the Beijing Genomics Institute and, you know, they're very well respected . . . but maybe some of our Scottish people might think, "Our samples are going to China? What's that about? What about the cleaner? What do they get paid? What's their trade union right, you know?" So, I actually think there would be some people that if you actually asked them the question, "We want to send your sample to China, how would you feel about that?" they might not be thrilled. Now . . . I don't want to get straight into political incorrectness and I am not making a huge point, but I do think you got to think about what's the extreme here. You know if it's Stanford University nobody's going to blink.[90]

Stanford, but what about England? As a colleague responded: "Scotland has a very strong national identity and, you know, I think that your point about samples going to China might well be true for certain individuals, but equally there are certain individuals that I am quite sure in Scotland would say, 'My samples went to England, good lord!'"[91] Is Scotland part of the same country as England? For GS, formed during the era of devolution, this was a relevant and important question. The commitment to develop the health and wealth of the people of Scotland had been core to the initial GS vision. Yet as GS moved into the postgenomic era, one in which networked systems of rapid technological change mattered more than bounded efforts to understand discrete populations, how could this commitment be maintained?

If only the scientific and technical considerations could be separated from the ethical ones, the answer seemed obvious: "The main place in the UK that's doing really high throughput sequencing is the Sanger

Institute, and it would be bonkers for us not to be able to send samples there because we are not trying to compete with them."[92] However, no such separation proved possible. GS had informed many of the participants through their information leaflets that only Scottish researchers would study the samples. Thus, despite the technical advantages of sending the samples to England, to do so would put the integrity of the informed consent process in jeopardy.

In the face of this dilemma, those in charge of the practical implementation of GS went back to the social scientists who had conducted the public consultations, and who they considered experts on the views and preferences of the Scottish people. However, by this time it had been several years since the social scientists had actively worked on GS. Support for their public consultation and engagement work had not been included in the second round of GS funding from the Scottish government. The global financial crisis had hit, and the Scottish government, like governments around the world, slashed research budgets. In the wake of these changes, GS sought funding only for those activities it deemed most vital, and public engagement activities were not among these.[93] Thus, when the question of samples leaving the country came up, the social science infrastructure for understanding public views no longer existed.

With social scientists no longer actively engaged, GS leadership turned to UK research councils and ethics bodies to advise them. These bodies concluded that the biological samples could be sent to England. As one organizer explained: "Although we'd said initially Scotland, because it's a national health service, although Scotland's a nation, it's still part of the UK and it was felt that actually sending them within the UK was absolutely fine."[94]

Sending samples to other places in the United Kingdom was deemed acceptable, but what about to Stanford University? Could the samples go to California?

For this question, it was possible to go back to the public survey data. While not much relevant data had been collected on sharing the GS resource outside of Scotland—for at the time all thought the samples would stay in the country—some data did exist on this topic. Specifically, the public survey of 1,001 Scottish residents indicated that a significant minority (28 percent) did not want researchers outside of the United Kingdom to access GS data. One might then infer that if this many did not want non-UK researchers to have access to data, perhaps more would have concerns about the actual biological samples.[95]

When I asked GS organizers how they thought about this issue of whether Scottish DNA could leave the country, this survey data did not come up. Instead, they explained that they went back to the NHS research ethics service, but this time NHS had no advice: "It was really challenging for us because we went to our ethics committee and they said well we only deal with the ethics of stuff in the UK so we've no opinion on this."[96] This governance vacuum created a challenge for GS and brings into view the more fundamental postgenomic questions of knowledge and justice they faced.

Of Mutants and Machines

Researchers asked Generation Scotland for permission to send samples to California to help solve a hard, technical problem: the human mutation rate between generations. Finding mutations between generations of human beings meant finding minute differences: maybe only one or two changes in the three billion nucleotides that make up a human genome. Thus, it is a hard problem that if solved would push sequencing technologies forward. GS was well positioned to help, as it had collected samples from multiple generations. Sanger, home of the best sequencing machines in the United Kingdom, had been cleared to sequence the samples. Although the next step would be to send them on to a biotech company in Silicon Valley so that the accuracy of the Sanger sequence could be tested with these new machines, the GS leadership felt they could not send the samples: "The Sanger Institute . . . they've done their thing, but because they are at . . . the real leading edge of sequencing technology . . . they collaborate with other centers who are equally pushing the boundaries, and so there is a biotech company in California that has developed a really quite innovative new method for sequencing and they've offered free of charge to resequence the same samples to have a direct comparison of the two technologies. Now that's to GS's benefit, it's to Sanger's benefit, it's to researcher's benefit, but we can't send the samples."[97] This example is notable for a number of reasons. It is not just that GS wanted to send the samples out of Scotland, but also that they wanted to do so to contribute to efforts to develop better sequencing technology and machines. The goal was not, as it had been in the early years of GS, to advance the health of the people of Scotland. By definition, de novo mutations are not ones that can be predicted, and it is hard to see a way in which knowing that one's ailments are caused by mutations that originated with oneself would improve one's health. This is not to say that the question is not biologically interesting, or that it might not eventually produce medically important insights.[98] It is only to say that technological advance, not medical advance, was the primary goal.

However, this attention to developing the capacities of the machines, and not the health of the Scottish people, did not represent a turn from questions of ethics. Far from it. The GS team remained deeply committed to creating an ethical approach. What changed was the nature of the ethical questions and the methods used to address them. Rather than possess a thing of value that they sought to mine for the good of the health and wealth of Scotland, GS found themselves attempting to create something of value. The best way to create this value, they argued, was by using the resource.

But this led back again to the question, to whom should this value accrue? In the early years of GS, the answer was clear: it should be valuable for Scottish people. However, as the GS team sought to rework the access policy of GS so that samples could be sent outside of the country, the ethical valence of serving the Scottish people shifted. Rather than the right thing to do, it became a problem. As one GS leader explained to me in the summer of 2012: "Emphasizing the Scottishness has turned out to be detrimental in a way because it's inhibiting the potential use of the samples. It's potentially affecting the existence—i.e., sustainability—of Generation Scotland. And if you look at making the best use of these donated samples as being the most ethical course of action, then it's inhibiting that."[99] Rather than an enlightened approach to government, the attempt to create a resource of benefit to the Scottish public is recast as an attempt by the Scottish government to create economic gains. As one GS organizer explained: "What they [the Scottish government] wanted was for the research to be done in Scotland, to the benefit of Scotland, Scottish people, and to highlight Scotland as a technological, innovative medical research hub."[100] In retrospect, the approach had limits:

> A bit of kind of Scottishness crept in, I think for good reasons, to kind of make people feel it was a Scottish project. . . . But one . . . unintended side effect of that was that people were actually told the sam-

> ples would stay in Scotland and would be researched on by Scottish researchers. And of course that starts to get a bit silly because although Scotland has some great researchers, it is only less than 10 percent of the UK, and although again we have got some great technology up here, there are places like the Sanger Institute in Cambridge which is the UK leader in genomic technology.[101]

It seemed plainly evident that after all the time, effort, and money that had been invested by the Scottish government, researchers, and people into GS, the only ethical course would be to use and develop the GS resource, and this meant sending the samples outside of Scotland. Thus, at the end of summer 2012 GS leadership decided to recontact all 24,000 participants in the Scottish Family Health Study to ask for reconsent to allow the samples to travel outside the country.[102]

Postgenomic Shifts: From Governing People to Ethical Humans

Generation Scotland's changed position reflects the substantial shifts that had occurred since GS leaders first proposed the initiative in the late 1990s. When ideas for Generation Scotland first formed, not only was it possible to believe that genomes had great biological and economic value, it was also possible to assume that these genomes could and should be bound by national borders. Far from exploitation, this represented an act of democratic sovereignty.[103] A decade later, these assumptions of the genome era no longer proved workable. Genomes, like many other things in the wake of the financial crisis, had dropped in value. To regenerate, they would, like capital, need to travel across borders and participate in global flows. These changed conditions changed the questions of knowledge and pressed ethics into the constitutional questions of justice.

Because knowledge and the application of it to medical problems would not be immediately forthcoming, genomics now posed questions about how to move from a deluge of sequencing data to relevant medical insights. This change in the problem of knowledge carried along with it a change in the problem of ethics. At the beginning of GS, when many considered genomes highly valuable, equitable distribution of this value appeared the pressing ethical issue. Thus countries such as Scotland and Mexico sought to protect and develop their own genomes, rather than allow more powerful nations to appropriate them. However, the postgenomic fall in value of genomes made this issue less salient. Rather than fairly distributing value, the problem became how to constitute value of genomes in the first place.

It is in this context that a cosmopolitan ethic of care began to seem more important than a liberal democratic right of control. At the time, the notion of cosmopolitanism was gaining traction on the international stage. Prominent scholars such as Paul Gilroy, Anthony Kwame Appiah, and Judith Butler recently had published books that called citizens around the world to care for those who lived beyond their borders. Such a cosmopolitan ethics, they argued, would foster the constitution of twenty-first-century nonviolent global societies.[104] Around the same time, some GS leaders began to argue that this cosmopolitan ethics supported not just global societies, but global sciences. It provided the basis for citizens in one country to share their genomic data with those in another, an act of sharing that would foster aggregating genomes in numbers high enough to enable meaningful genomic analysis.[105] Referencing Framingham, the town in the United States in which residents gave samples for the famous long-term epidemiological study of cardiovascular disease, a leader of GS explained:

> I think that Framingham's proud of the fact that they've contributed to an understanding of heart disease at a global level and would not—I can't imagine for a moment that they would've wanted that knowledge and information to be kept to Framingham, or even to North America, but realize and see its relevance to Europe and Asia and Africa. I would argue that that's what we should be aiming for in Generation Scotland, and I'm sure that that's what the Scots that contribute to it would want to see happen. They would want to see things that we discover through that study become widely seen, used and adopted.[106]

This new ethical framing significantly reworked the liberal democratic vision—in particular, the commitment to public sovereignty—that so strongly shaped the original GS vision. Rather than a project aimed at constructing and using a resource that could help the people of Scotland, this appeal to a cosmopolitan ethics emphasized the role that GS and the people of Scotland might play in advancing science and helping all humans around the globe. This vision of GS was not grounded in an empirical study of the particular views of the people of Scotland, but in an ideal of human altruism.[107] It framed the advancement of science and its machines as the way forward for all humans.

For Whom Should Our Blood Flow? Democratic People, Cosmopolitan Humans, or the Machine?

Rather than a democratic science, a science of and for the people of Scotland, GS now offers a global vision of the advancement of humanity. Rather than just the Scots, it seeks benefits for all the world's people. Yet, for now, the beneficiaries of access to Scottish DNA may only be technology entrepreneurs and their machines. This practical reality raises a more fundamental question: Should science and technology advance, even in the absence of clear broader benefits? This question is raised by the GS case, as it has been by many other human pursuits of science and technology. However, equally, GS poses a more unique question: might democracy prevail in the absence of any clear benefits to humans?

Consider why in the first instance so much blood was collected from the people of Scotland from 2006 to 2011. Arguably, hospitals and blood banks already collect enough blood to provide all the DNA needed for genomic research. So why did GS set out to collect so much more new blood? While there are many answers, perhaps the most salient is the widely held liberal democratic commitment to the right to control one's own body. Many in Euro-American liberal democracies believe, like Arendt, that the body is a personal, intimate space that should not be subjected to "the full light" of public scrutiny without informed consent.[108] Both—autonomy over one's body and informed consent—are pillars of modern Euro-American liberal democracies. So, blood can be given, but only with consent, and only for discrete acts known to consenting individuals. Thus, GS collected blood anew. In other words, it generated so much blood partly because there is so much democracy.

These biopolitical logics suited the era of "the human genome," an era in which both blood and genomes harbored great meaning for many. Today, however, as we are increasingly told that our genomes by themselves mean little, and only in aggregate reveal something, it is less clear how we might appropriately ask individuals to make sense of genomic research in order that they might help govern it. To many, the human genome sequence did appear to be a thing one could know and control.

Indeed, so much so that it inspired dystopias premised on humans becoming God. But what came after the human genome has been in no way a well-bounded thing. It is not so much information, as "in-formation."

GS leaders faced the dilemmas of responding to these conditions with liberal democratic instruments and assumptions. They presumed they could know and inform the people of Scotland about what came after the genome: a world of valuable genomes that could be mined with technology already existing in Scotland. Unlike in Iceland, in Scotland these presumptions and practices did ultimately lead to the creation of a national collection of DNA samples and personal data. However, Scotland today has created a collection of blood, genomes, and urine with unclear value both for the people of Scotland and researchers. The uncertainty of its value in part derives from the liberal democratic demand to gain consent to send the DNA outside of the United Kingdom. As of February 13, 2015, 11,255 of GS participants agreed to give consent.[109] This is a little less than half of the 24,000 original participants, a number that reduces the power of the collection.

Today, in Scotland, the GS team is attempting to address these problems. When I traveled there in February 2013, some of the original team of social scientists and lawyers and I gathered to discuss the problems of informed consent, and the challenge of governing entities like GS whose form shifts over time and whose value is uncertain and not known in advance. These are ongoing challenges, and ones that demand increased attention from scientists, social scientists, and lawyers.

And yet despite these limits of liberal democratic concepts and practices, their appeal is obvious. They tap into a deeply rooted Enlightenment belief that the world is made up of things about which we can know and gather around to make decisions. We gather because there are things of concern and value. It is only appropriate that in the genomic era it would be the Althing in Iceland—the oldest extant parliament in the world—that would first constitute the genome as a thing of national importance. But in these postgenomic times, is there any "thing" at all? If not, and we are unable to gather around a thing of clear value, should scientists and society proceed?

These are the questions that define the postgenomic conditions in which today Generation Scotland and other efforts to harness genomics to biomedical and societal value seek a way forward. Soon after I returned from Scotland in the spring of 2013, a front-page article in the

New York Times reporting on "the race" to sequence cancer patient genomes concluded with the following observation from Dr. James M. Crawford, chair of pathology at Hofstra North Shore–LIJ School of Medicine: "What is the ultimate utility of this personalized [genomic] medicine? . . . As a medical profession but also as a society we have not answered this question to our satisfaction." And so, for now, Dr. Crawford's institution remains "quite literally on the fence" about whether to invest.[110]

But maybe someone would make something of nothing. In the United States, this is the fabled dream. And it would be in the United States that the next major effort to make genomes meaningful for people would arise. Genomes would indeed travel to California, and this time the corporate stakes would be clear.

6 GENOMICS FOR THE 98 PERCENT?

We'd like to reach 98pc of the world, that is our goal.
—Anne Wojcicki, co-founder of 23andMe, at the World Economic Forum

While geneticists and government leaders in Scotland navigated the beliefs and practices of nations, across the world in Silicon Valley informatic entrepreneurs launched efforts to reconstitute the bases of social and scientific life. Scotland collected its first DNA samples in 2006. In that same year, Apple placed the Internet in people's pockets with the introduction of the iPhone, and Facebook opened up its social networking site to anyone over age thirteen. Promises that these expanded powers of communication and information would radically open up societies circulated widely, generating much enthusiasm and capital investment. With the power to connect people and information now at everyone's fingertips, the leaders of this digital informatic revolution—dubbed by some the digerati—promised that long-entrenched institutions—from federal governments to Wall Street–backed large corporations—no longer would dominate. Revolutionary forms of capital, commodities, and community would flourish.[1]

And science. In November 2007, just a year into the era of social networking, founders of the Google-backed start-up company 23andMe set out to articulate genomics to social networking platforms. Like personal computers, 23andMe predicted the resulting personal genomics (PG) would rapidly move from an innovation with unclear value to one that was indispensable to contemporary life. The key to this transformation would be to untether genomics from the rigid ways of bureaucracies and nations and to empower people to directly control their genomic data. This, the company argued, would lead to nothing less than a revolution—for both the life sciences, and democracy.

23andMe was but one of several personal genomics companies launched between 2007 and 2009.[2] I focus on 23andMe in this chapter as it is the only PG company to survive the initial fanfare, and it is the one that most clearly built its identity around the aspiration of democratization. In the fall of 2008, when 23andMe dropped its cost from $999 to $399, its website proudly announced: 23andMe Democratizes Personal Genetics.[3] The following year, co-founder Linda Avey explained that the main reason why she and Anne Wojcicki co-founded the company was to "take genetics out of the protective realm of the scientific community and make it accessible to the lay public."[4]

Yet where some saw liberation, others saw exploitation. In 2006, foreshadowing the debate to come, Gordon H. Smith, chairman of the congressional committee that oversaw the 2006 hearing on home DNA testing, called direct-to-consumer genetic testing "modern-day snake oil."[5] In June 2008, the California Department of Public Health sent thirteen genetic testing companies, including 23andMe, "cease and desist letters," charging them with violating state law by offering genetic tests without a physician's order.[6] Two years later, in June 2010, the US Food and Drug Administration (FDA) sent letters to 23andMe and four other PG companies informing them that they must receive regulatory approval for their services. The FDA, which up until then had refrained from regulating genetic tests, cited worries about analytical and clinical accuracy.[7] On November 22, 2013, after not hearing from 23andMe for six months and again citing concerns about the accuracy of its tests, the FDA ordered 23andMe to stop marketing its personal genome service. Five days later, on November 27, 2013, San Diego resident Lisa Casey filed a class action suit against 23andMe in the Southern District Court of California. The lawsuit stated simply: "The test results are meaningless."[8]

What had started in 2007 as a story of David versus Goliath—23andMe, a small start-up company, attempts to overthrow unjust control by the powerful state—by 2013 had turned into the familiar story of the corporate corruption of science. Yet such clear stories of good and bad, heroes and villains, fail to bring into view the more interesting and important dynamics in play. As I seek to demonstrate in the pages that follow, the debates over personal genomics make visible a broader contemporary struggle over how to constitute knowledge and justice in the midst of challenges to the credibility of dominant institutions, and investments in informatics as the new infrastructure for collective living and understanding. Both the state (for example, the California Department of Public Health and the US Federal Drug Administration) and companies (for example, 23andMe) accused one another of unjust interference in the democratic flow of information required for civic understanding. The state charged companies with failing to live up to important government standards that ensured the accurate transmission of information to citizens. Companies charged the state with blocking access to information. Both adhered passionately to their positions.

In so doing, both sides failed to address the more fundamental question: Can *any* person—including biological and medical experts—interpret genomic data in a manner that produces valuable knowledge for scientific and social life? Or is genomics moving the life sciences into the unstable epistemic and social space that Arendt described, a space in which scientists and nonscientists alike are unable "to think what we are doing."[9] Advocates of personal genomics argue that the field's interpretive challenges will be solved through amassing more data and making it accessible to more and more people. Yet this argument sits uneasily with the on-the-ground realities. As we will see, making genomic data valuable is a very hard problem, even when DNA can be sequenced on the latest sequencing machines produced in Silicon Valley, and the data can be analyzed using the most powerful algorithms. Efforts to make sense of it entail computer clouds, algorithms, and bioinformatics on a scale that has attracted the interest of Google.[10] They also sometimes require the judgment of human beings. How to constitute right relations between the machines and these human beings is far from a settled matter.

To bring these more fundamental questions about how to know life in the midst of a growing sea of genomic data, I draw from my experiences living amid the rise of the personal genomics industry over the last decade. In January 2006, I moved from the East Coast to take up a position at University of California, Santa Cruz (UCSC). UCSC, as explained earlier in the book, is known for its work creating the bioinformatics tools needed to make genomic data available and useful to researchers. Thus, not surprisingly, companies like 23andMe recruited some of the researchers I came to know during my first years there. Over the years,

I benefited from being situated between UCSC and these companies based "over the hill" in Silicon Valley. I met and was able to interview bioinformaticians who worked both in the academy and in the emerging industry (with many spending time in both arenas), visited personal genetic and related bioinformatics companies located in Silicon Valley, attended seminars given by company leaders, and helped to organize ones that gathered academic and industry practitioners.[11] This chapter draws upon these experiences to move beyond the story of corporate corruption of science oft told about this new industry to forge new stories that open up the more fundamental questions of knowledge and justice at play.[12]

Rather than threatening the integrity of scientific or democratic practice, through yoking the locus of agency in liberal democracies—the person—to what many hope and believe is the locus of agency in the life sciences—genomes—personal genomics created a potent zone of biosocial formation.[13] 23andMe presented a clear and powerful vision of how to make good on this powerful convergence: place people—not scientific experts or government regulators—at the heart of the action; give them access to all information—including their genomic information. It argued that empowering people to use informatics tools to know their own bodies would lead to breakthroughs in science, medicine, and democracy. However, trouble quickly arose as the route between genomic information and the hoped-for biological knowledge and democratic power proved much less clear than 23andMe envisioned. Rather than opening up access to valuable genomic data, like Generation Scotland, personal genomics companies faced the problem of how in the first instance to create valuable knowledge from genomic data.

Genomics Gone Google

In the fall of 2007, with the support of Google co-founder and billionaire Sergey Brin, 23andMe was among the first to offer customers direct access to their raw genomic data and interpretations of its medical significance. Brin and 23andMe co-founder Linda Avey met the previous year through Brin's interest in Parkinson's disease. Parkinson's afflicted his mother and associated with the *LRRK2* genetic variant, a variant contained within his own genome. Avey worked at Perlegen, a company that did much of the early single nucleotide polymorphism (SNP) discovery in the human genome and conducted the first genome-wide association studies (GWAS). Shortly before Avey and Brin met, Perlegen partnered with the Mayo Clinic and the Michael J. Fox Foundation to study Parkinson's disease through comparing the genomes of those with Parkinson's (the cases) to those without Parkinson's (the controls). The goal was to determine if there were any genetic variations that could be associated with the disease. However, the study suffered from low numbers. Perlegen had access to only three hundred DNA samples from people with Parkinson's, too few to create statistically robust associations. Reportedly, Brin immediately honed into these mathematical challenges, asking Avey what *P* values (a measure of statistical significance) Perlegen used to determine genetic associations.

Not long after their meeting, Avey left Perlegen but kept in touch with Brin and became increasingly interested in the intersection between the Google world of data management and the data-rich world of genomics. What if, Avey wondered, one could Google their own genome? If millions of people could do this, it would solve the problem of small numbers faced in the Parkinson's study. Anne Wojcicki, who Brin married the following year, found the idea compelling. Although at the time a hedge fund investor, Wojcicki held a bachelor's degree in biology. Taken with Avey's vision of googling genomes, Wojcicki quit her job and teamed up with Avey.

Powered by Google's money and algorithmic prowess, Avey and Wojcicki sought to turn assumptions about genetic information on their head.[14] Why assume, as the dominant approach to federally funded human genetics research did, that genetics was dangerous and harmful, and that people should be protected from it? This, after all, had led to a dreadful state of affairs, one Avey described as "feudal," in which geneticist overlords "protected" their research subjects from genetic information.[15] Why not assume instead, as Google did, that information is good; that information tailored to individual needs and wants is even better; that sharing this personal information is even better still? They set out to establish the informatic infrastructure that would make this, the people's genomic revolution, possible.

Specifically, they created spit kits. Anyone could purchase these kits and send in their DNA-rich spit for genotyping.[16] Customers could then browse their results on a fun, hip, Crayola-colored interactive website and share them with friends and family.[17]

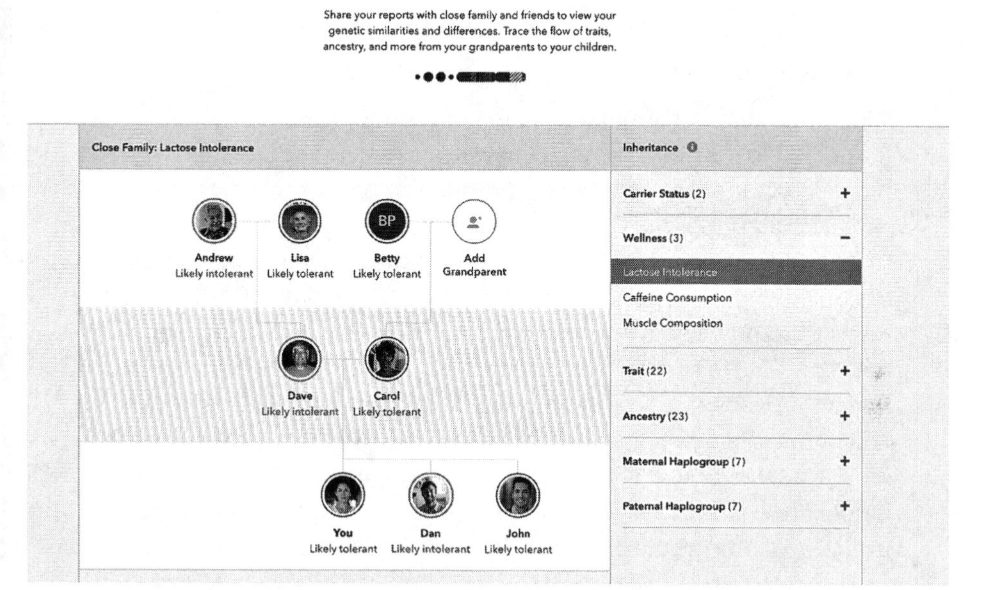

Harnessing the power of social media to the consumer's right to know, they sought to transform the passive objectified subjects of human genetics into the empowered people of genomics 2.0. They announced this relationship they intended to forge between the human body's twenty-three chromosomes and everyday people in the name they chose for the company: 23andMe. No longer would those twenty-three chromosomes be the domain of scientific experts. As Linda Avey explained in her inaugural post to her personal blogsite, *The Life and Times of Lilly Mendel*: "I still strongly believe in the main reason why my co-founder, Anne Wojcicki, and I started 23andMe in the first place—to take genetics out of the protective realm of the scientific community and make it accessible to *the lay public*. . . . [V]ery vocal scientists seem to be quite threatened by this notion of *democratizing* DNA" (italics added).[18] Company leaders reported encountering academic scientists who viewed their efforts very negatively, accusing 23andMe of "belittling and trivializing the science."[19] It is this attitude that 23andMe sought to change. The company, its leaders argued, would ditch human genetics' historic paternalism and feudalism in favor of empowerment and democracy.

A Research Revolution?

While 23andMe foregrounded these political aspirations, ultimately its founders and directors sought to revolutionize genomic research by building the largest database in the world that linked genomic data with phenotypic data (for example, data about diet, medical history, and such).[20] In early 2008, 23andMe director and technology investor Esther Dyson celebrated these aspirations to create a "large database of genetic information" and to use it for research. She asked 23andMe customers to take part not "for purely altruistic reasons," but because it could lead to research results that "could translate directly into benefits for you, or at least for your children, grandchildren, and friends."[21] 23andMe called this endeavor 23andWe and invited its customers to join this "research revolution."[22] An airship that at times flew directly in front of my own home invited all the residents of the Bay Area to "Join the Research Revolution."

Although Dyson described this research vision as 23andMe's "second goal," for Brin and Wojcicki, advancing research always had been the motivating force. If advances in understanding the genetics of Parkinson's disease did not help Brin, Brin and Wojcicki hoped it might help their son Benji, who was born shortly after the launch of 23andWe. Thus, not surprisingly, 23andWe launched with a large-scale study of Parkinson's disease. The aim was to recruit ten thousand research subjects, thirty times more than the Perlegen study. To achieve this ambitious goal, 23andMe opened the study up to anyone diagnosed with Parkinson's who also had access to the Internet.[23] With no geographic borders and no doctors to limit entrance to the study, the leaders of 23andMe believed they could greatly expand their recruitment power.[24] Other 23andMe customers who did not have the disease could participate through serving as controls. Brin provided most of the funding for the study as well as his own DNA.[25]

The Parkinson's study was emblematic of the 23andMe vision. It brought social media strategies to the field of human genomics in order to increase exponentially the number of research subjects in human genomics as well as the data one could gather about them. While research was serious business, 23andMe sought to create a web experience that was fun and interactive so that their customers would want to come back to the web portal over and over again to check on updates as well as their "To Do" box. The company used this To Do box to send customers surveys designed to collect their phenotypic data. Upon completion, an automated message thanked customers and told them their information would help to "customize their experience" and, if they had consented to 23andWe, it would "contribute anonymously to this groundbreaking effort."[26]

The ability to use social media strategies to build ongoing relationships with their customers was the real power of 23andMe. Genome scientists knew how to collect DNA and how to genotype it. What they lacked was the capacity to create the ongoing relationships with subjects required to collect the phenotypic data needed to make sense of the genomic data. 23andMe, as one of its data scientists explained to me, offered a solution: "The genetic piece is the easy part. The phenotype is the tough part. . . . I think scientifically the value that we have . . . is the phenotype side. The phenotypic information we're gathering is very uniform. Everybody answers the same survey, and this information that we're laying on top of the genomics is, I think, pretty unprecedented. That's our use of the Web: to put surveys up to everyone and say, 'Hey, you guys. Answer this one.'"[27] 23andMe believed its customers would respond to their calls to complete the surveys because of the connections the company forged between chromosomes and personal identity. Through making genomics interactive and participatory—by enabling customers to learn in an ongoing way in real time along with the company about their genomes—23andMe sought to make genomics of personal interest—as personal as the personal computer. A *personal genomics*. Through harnessing genomics to the interests and power of persons, 23andMe sought to revolutionize the field. A few academic scientist overlords, they asserted, would no longer rule genomics; instead, the people would.[28]

Informatics and Freedom: A Contested Past

While a powerful and charismatic vision, articulating a new form of informatics—this time genomics—to personal identity and freedom has never been an easy task. Informatics and freedom have a long and contested past. Early in the twentieth century, computers were primarily tools of warfare. For many who lived during the last century, they represented antidemocratic, authoritarian tendencies. As social media historian Fred Turner chronicled in his book *From Counterculture to Cyberculture: Stewart Brand, the Whole Earth Network, and the Rise of Digital Utopianism*, in the 1960s free speech activists wore computer cards as signs of protest. Mario Savio, author of the founding manifesto of the free speech movement, believed that computers and punch cards were turning human beings into automated cogs in a technological society. Others contested Savio's dystopic vision of informatics and human freedom. Most notably, in the late 1960s Stewart Brand and the Whole Earth network forged connections between the counterculture movement and Silicon Valley. They viewed informatics not as the tool of the oppressor but of self-sustaining lifestyles.[29]

While the tension between these two contrasting images of digital technologies—as domination or liberation—continues to this day, the latter grew in strength. Apple may have lost much of its credibility as the anti-IBM, anticorporation that creates technologies that promote human freedoms; however, the whole line of so-called personal technologies it sells—from the iMac to the iPod to the iPad—only has grown in popularity. 23andMe hoped human genomics would undergo a similar transformation, losing its associations with authoritarianism and gaining a new identity as the latest hip piece of personal technology.

Yet forty years on from the launch of the Do-It-Yourself (DIY) movement popularized by Brand's *Whole Earth Catalog*, DIY genetics faced a different, and in some ways more difficult, set of challenges. It was one thing to participate in what Jodi Dean calls communicative capitalism and be a consumer in-waiting as you play the free version of Angry Birds on your iPhone (for how then could you resist purchasing the full version for $2.99?). However, it was quite another to partake in biocapitalism and to become a consumer and patient-in-waiting as you pay to play with your genome on the 23andMe website. As Kaushik Sunder Rajan explains, biocapital is "not just the encroachment of capital on a new domain of the life science," it is a fundamental reworking of capital and the life sciences.[30] 23andMe provides an exemplary case. The company does not aspire to just make money by implementing an already existing approach to the study of life. Rather, it aims to revolutionize the very study of life through creating a new corporate approach. These aspirations created significant challenges—as well as attention—from the very moment 23andMe opened its doors in 2007.

To be clear, difficulties did not arise because 23andMe represented a private effort to revolutionize a field of science. After all, in the fields of information and communication technologies companies have long led the way. For decades Bell Labs attracted the best and brightest scientists and engineers and innovated some of the twentieth century's most important forms of science and technology: transistors, lasers, information theory.[31] Bell Labs is not the only example. As an early employee of 23andMe explained to me: "You go to the computer science department at Stanford and they love Intel. Intel's innovating all the time and bringing them better stuff to work on." However, in the field of health it is different: "It's the health piece of it that puts the knot in, I think, and it's really unfortunate. . . . We're talking about people's lives and how they stay alive . . . and companies coming in profiteering in that space just makes people feel a little queasy."[32] It also is important to recall that early on genome scientists sought to define their field as an open, public science that opposed commercialization of genomic information. Like Celera, for many 23andMe represented a threat to long-standing moral economies that operated based on the principles of openness and data sharing.[33]

23andMe also differed from other successful social media companies by charging their customers from the start, and the cost was substantial. "People were paying to sign up, which is sort of the opposite of the Googles and the Twitters and Facebooks of the world where everything's free. So, we were very contrary to that."[34] When the company opened in 2007, sequencing a full human genome still cost millions of dollars. 23andMe's genotyping of six hundred thousand of the three billion nucleotides that make up a full human genome could not yet be done cheaply, and the company needed to at least break even. As one member of the early management team explained to me: "We couldn't be losing tons of money as a start-up."[35] Yet charging a thousand dollars when others were writing manifestos about informatics' free economy raised eyebrows.[36] Was 23andMe worth it?

Not only did critics worry that consumers might be overcharged, they also believed that 23andMe put their customers at risk through opening up new pathways for genetic discrimination. This time it might not

23andMe sticker picked up by author at 23andMe headquarters in Mountain View, California, in June 2010.

be the state, but a multinational corporation leading society down this dangerous path.[37] While company leaders argued that it was "not part of Google by any stretch of the imagination," Google funding and Brin's personal interest forged a strong tie between the informatics giant and 23andMe.[38]

Avey and Wojcicki forged their careers during the rise of digital utopianism and lived at its epicenter: Silicon Valley. They believed in its imaginaries of connection and convergence made possible by the digital world's universal language of zeros and ones.[39] They believed that companies could do good and live by Google's injunction: "Don't be evil."[40] However, as 23andMe quickly discovered, these beliefs do not translate seamlessly to the world of biomedicine. While in fun media forms such as stickers, 23andMe offered engaging and appealing images of what the people and genomics of personal genomics might look like, in practice who these people might be, and what they could do with genomes remained far from clear.

Who Are the Persons of Personal Genomics?

In its effort to place people, and not populations or nations, at the center of genomic research, 23andMe set out into largely unchartered scientific and ethical waters. While George Church at Harvard University announced in 2005 the Personal Genome Project (PGP)—about which we will learn more in the next chapter—it would be 23andMe that broadly publicized the idea that people could and should link to their genomic data.[41] This so-called personal approach represented a major departure.

Up until 23andMe and the PGP, the leaders of genomic research had astutely avoided making people the subjects of research.[42] Instead, motivated by fears that genetic information could reveal life-altering information that might produce anxiety, stigma, and discrimination, they had placed considerable time and resources into creating systems that ensured that the individuals who provided DNA could not be linked back to their genetic information. Even in the case of the HapMap, where researchers did engage their subjects in the design and regulation of research, great care was taken to ensure that the only identifying information on the HapMap samples was population identifiers.

Yet as the HapMap ended, and more and more researchers sought to scan human genomes for variations associated with traits and diseases, maintaining firewalls between a person's genome and their personal identifying information began to look untenable. Several human genomics researchers began to worry that if the NHGRI and other major funding bodies maintained this strict approach to privacy the field would soon hit a wall. Fundamental new biological knowledge about human beings simply could not be created without wide-scale linkage of genotypic and phenotypic information. The field needed to find people whose genomes they could sequence and link to identifying phenotypic information.

Initially, they turned to those in their own ranks. Craig Venter led the way. He provided DNA for the Celera human genome sequence published in 2000 as well as the first diploid human genome published in 2007.[43] James Watson, co-discoverer of the double helix and first director of the Human Genome Project, provided the company 454 with his DNA and became the second person to have their genome fully sequenced. Some dismissed these efforts as vanity projects. Others called them "tacky."[44] Many questioned their scientific value. Even Watson himself believed sequencing his genome would reveal little of significance: "I realised one reason why I so easily said that I'd look at my genome is that I'm seventy-nine and one of the genes could have put me at risk of going blind at fifty. So we don't know enough about the information. Until thousands of people are sequenced, we won't really know, in most cases, what they mean."[45] But who would those thousands of people be? How could sequencing one's genome move from the elite activity of a privileged few to a respected activity of the many? 23andMe offered an answer: enroll subjects who understood genomics and who might already be inclined to take risks.

Genomics for Sky Divers and the Digerati

Among those the company originally targeted were skydivers. Indeed, reportedly 23andMe received a positive response from a society representing them in central California.[46] However, mostly the company sought out the so-called digerati. As one early employee at 23andMe explained to me: "The way we got the word out was very Internet-based and viral . . . [through] knowing who the nodes are in the digerati social networks."[47] The cost of the service also acted to attract the "right" kind of consumer. Initially, 23andMe cost $999.[48] Although early critics noted that this high price excluded many, it also helped the company target "a certain demographic that is more likely to be educated."[49]

In its opening year, 23andMe and other personal genomics companies courted this upper-class demographic. Navigenics, 23andMe's main competitor, held a bash in lower Manhattan in April 2008. In September 2008, 23andMe threw a "spit party" in the Chelsea district of New York City as part of New York Fashion Week. The likes of the Murdochs and Trumps, as well as an assortment of high-end fashion designers, attended. In between drinking cocktails, they spat their DNA into tubes. As reported in the *New York Times* Sunday Style section: "Co-founded by Anne Wojcicki, the wife of a founder of Google, the company, which has token financial backing from Harvey Weinstein and Wendi Murdoch, hopes to make spitting into a test tube as stylish as ordering a ginger martini."[50] A spit party as a mark of upper-class distinction?[51] As a scientist close to the industry explained: "At the original price especially it was certainly an item of conspicuous consumption. . . . It's like, 'Oh, I've had my genome scanned!' "[52]

23andMe quickly became the chic new thing. Wojcicki and Avey appeared on *Oprah*, the *Today Show*, and in the pages of *Fortune* and *Forbes* magazines. Yet ultimately these elite circles did not support the company's bigger vision. The digerati and the rich were an exclusive and relatively homogenous group. Almost all self-identified as white. Access to only their DNA limited 23andMe's market and research potential, something the company understood early on presented a problem.[53] As an early strategist for 23andMe explained: "We've always been talking about how do we improve the diversity in our database . . . and it really came from the business side. You know, we have people with South Asian backgrounds or Chinese backgrounds coming in and saying, 'This is bologna. None of this applies to me.' They'd be angry and I don't blame them."[54] While the digerati supported the launch of 23andMe, to achieve its goals the company needed to expand its customer base. At the World Economic Forum held in Davos, Switzerland, in January 2008, Wojcicki set her eyes on the whole globe: "We'd like to reach 98pc of the world, that is our goal."[55]

Focusing on the digerati and the rich and famous also did not work politically. To change the prevailing ethos and open up access to genomic data, 23andMe needed a strong moral argument. They found it in harnessing their vision of personal genomics to aspirations of justice and democracy. Genomics, they argued, should be for all people, not just a few elite scientists. A blog post celebrating *Time* magazine naming 23andMe "Invention of the Year" articulated this vision: "We hope that everyone is as excited as we are to see not just our service recognized, but also our shared vision of personal genomics for all."[56] While a powerful vision, it also was one easily undercut by 23andMe's associations with the rich and powerful Googles and Murdochs of the world.

Constituting the 98 Percent

Thus, 23andMe set out to dramatically expand those who could count as in the know, and hence capable of taking part in genomics. One early tactic they pursued was to challenge prevailing bioethical constructions of who could be a subject of human genetics research. Specifically, they argued against the dominant view that the "lay public" did not have the skills needed to safely interact with genetic information. Argued Avey: "I think the research community has taken the notion of 'human subject protection' way too far, to the point of unchecked paternalism. . . . And I do think the lay public is capable of understanding that what is currently known about their DNA is mostly a work-in-progress."[57] Simultaneously, they created educational resources so that potential customers could learn about genomics. In a section of the 23andMe website called Genetics 101, cheerfully colored cartoon characters take viewers through the basics of genomics, explaining genomes, phenotypes, and single nucleotide variations.[58] In another video, customers receive a virtual lab tour and find out what happens to their spit kits once they reach the 23andMe labs.[59] Empowered with the information gained through these educational videos, 23andMe argues that anyone can understand the basics of genomics needed learn about their own genomes.

Further, its leadership argues that this capacity to know about one's own genome is a fundamental right. Writes Wojcicki: "I strongly disagree that one's genome is currently only useful in research and not for individual use. There are a number of highly useful genetic results that may be generated. . . . Is it right for Kaiser to tell me what information I can or cannot have about my own body and my own genes? I co-founded 23andMe, a personal genetics company, to enable individuals to access their genetic information—what we believe to be a fundamental right."[60] Genomic information, Wojcicki argued here and elsewhere, should no longer be the exclusive preserve of health care providers or genome scientists. This was not just a matter of science, but of justice. Celebrating the drop in the price of its service from $999 to $399, in the fall of 2008, the 23andMe website banner read: "23andMe Democratizes Personal Genetics."[61]

If the early efforts to sequence personal genomes were viewed as the preserve of the privileged few (for example, Venter and Watson), 23andMe sought just the opposite: "We wanted to be all encompassing as much as possible and that's where we felt being in the U.S. and California in particular [was important]. Look around! I mean there are people of every color, every background here."[62] 23andMe sought to make genomic information available to all, or at least 98 percent of all, people. This, they believed, was not just good politics, but good science and business.

Genomics for the Masses?

For regulators, however, this dramatic expansion of the people of personal genomics sounded an alarm. From their perspective, there was a large difference between marketing to the in-the-know *digerati* and *all* people. It might be possible to turn a blind eye when personal genomics was the domain of the former. However, in May 2010 when Pathway Genomics announced its plans to sell its genetic testing service to consumers from the shelves of US pharmacy giant Walgreens, personal genomics moved definitively out of the realm of the digerati and into the realm of mainstream. It also caught the attention of the FDA.[63] As FDA administrator Jeff Shuren explained during the summer 2010 congressional hearings on personal genomics, while the FDA already had been considering the issues raised by direct-to-consumer genetic testing, the Pathway Genomics effort to sell in Walgreens sent a clear sign that something needed to be done.[64]

The specter of the masses had been raised, sparking fears about what they might do once they accessed their genetic information off the shelf at their local pharmacy. Would they, as Congressman Phil Gingrey suggested, jump off a building? Or as FDA's Shuren argued, might they "make a decision that adversely affects their health, such as stopping or changing the dose of a medication or continuing an unhealthy lifestyle, without the intervention of a learned intermediary"?[65]

At play in these discussions were centuries-old debates about the threat of the masses to the proper functioning of a democracy, and the role of education and expertise in countering this threat. Dating back at least as far as the French and American revolutions, political philosophers such as Jean-Jacques Rousseau and Alexis de Tocqueville worried about the mob or mass rule: If the people should be, as John Locke argued, the locus of government, how can we ensure that they do not just pursue their individual interests instead of the common good?[66] Rousseau argued that it was the responsibility of government to educate its people so that they attained the level of rational discourse needed to rule.[67] Shuren clearly feared that without the help of education and educated experts (what he called "learned intermediaries"), the ill-informed masses who shop at Walgreens might act to harm themselves. These individual harms could become collective harms as they created new problems for an already overburdened American health care system.[68]

While Shuren suggested the answer might be for a "learned intermediary" to shepherd individuals through the genetic testing process, later he indicated the answer might not be so simple: "Some companies now are making claims about high-risk medical indications like cancer, about the likelihood of responding to a specific drug. In many cases, the link between the genetic results and the risk of developing a disease or drug response has not been well-established. *Even the experts don't know what the results mean*" (italics added).[69] Concerns that even experts do not know how to interpret genomic information arose frequently in the congressional hearings. Several members of Congress noted that different companies often provided customers with different results. Congressman Christensen, herself a doctor, asked if doctors can interpret the results and thus help consumers make sense of their personal genetics results. The author of the Government Accountability Office report, Gregory Kutz, answered: "The genetic experts we spoke to said that most doctors would not be able to interpret." Christensen responds: "So who can interpret it?" Later in the hearing, Jim Evans, medical geneticist and editor in chief of *Genetics in Medicine*, answered: "No one knows how to interpret these data. And that is quite clear."

Personal Genomics for Any Persons at All?

For those working in the field of personal genomics, this was not a new problem. Practitioners knew that creating genomic knowledge from the proliferating data was the challenge that lay ahead. Indeed, companies had moved into the space partly because there was no accepted approach to this problem, and there was much room for innovation. What did all the genomic data mean? Few answers existed. While this created the opportunity to innovate, it also created what a genome scientist at a leading personal genomics company described to me as a chicken-and-egg problem: "Well, that's kind of a chicken or egg problem. You need to have this industry develop, become mature enough so we can figure out what the right answer is. And the way to do that is to allow the early adopters to be able to play and fund it effectively."[70] Companies faced a dilemma: they promised genomic knowledge, but in many cases that knowledge did not exist. To produce it, they needed people who were willing to give up their DNA and dollars to help support the development of the field.

Early on, to solve this problem PG companies used an early adopter approach. For years, it had been a norm in web technology companies to beta test new technologies on potential consumers. The approach offered a mutual benefit: early adopters gained the distinction of being the first to try the latest gadget; companies received user feedback that allowed them to create products with greater value to consumers and companies alike. However, questions arose about whether genomics fit this model. Beginning in early 2008, prominent tech bloggers and scientists began to write articles noting disagreements among PG companies' test results.[71] In 2009, Craig Venter and colleagues aired these issues in the pages of the prominent journal *Nature*.[72] Many inside and outside the industry viewed these articles as damaging to the credibility of the field. Unlike an early adopter model in which the product evolves over time incorporating user values, these critiques held personal genomics to an ideal of objective science in which an invariant truth, not changeable values, guides the way.[73]

An Impersonal Personal Genomics

In an effort to respond to these critiques, industry leaders came together in the summer of 2008 to create standards. Coordinated by the Personal

Medicine Coalition (PMC), this effort included the three main companies in existence at the time: 23andMe, Navigenics, and deCODEme (a spin-off of deCODE). Company representatives who I spoke with in 2009 reported that all three made an honest and concerted effort to proactively work together and to put aside personal interests in order to create standardized approaches to genomic analysis that all could use. According to the PMC, their efforts were successful. The PMC acknowledged differences existed between the companies but characterized them as insignificant. In the end, they concluded that the "genetic risk profiles agree very well across the three companies when a similar number of validated SNP markers are used."[74]

A different picture of this effort to create industry standards emerged in interviews with the participants. While everyone I spoke with confirmed that all three companies shared a similar algorithm, they also agreed that such an algorithm produced different results if the companies fed different data into it. They considered questions about which data to use critical: Which "reads" of genomes produce data clean enough to find variants, or single nucleotide polymorphisms (SNPs)? Which SNPs should one pick to analyze? Which papers about these SNPs could be trusted to determine the odds of having a trait or developing a disease? While the exponential drop in the cost of sequencing made it possible to generate massive amounts of genomic data, these questions about how to make meaningful biological and medical discoveries from that data remained far from tractable.

While a problem, this space of uncertainty also created a business opportunity. There is too much data and analysis for anyone to process. Thus, company choices about which data and literature to focus on distinguished them. Choices embody values, and they create value. Genome sequence data is cheap and plentiful—too plentiful. What is expensive, and thus valuable, is the work of interpretation—decisions about what data to trust, and how to analyze it. As one molecular biologist and personal genomics practitioner explained:

> INT: People are wondering "Why isn't the price [of personal genomics services] going down if the price of genotyping itself is going down even faster than Moore's law for computer processing? Why isn't the price of personal genomics going down that fast?" The answer is that the data generation is

getting cheaper and cheaper, but the interpretation has not followed suit yet.

JR: It is not going down at the rate predicted by Moore's law?

INT: No, it is going up with time probably [laughter].[75]

What became clear in my interviews with practitioners in the field is that the work put into finding interesting and meaningful SNPs is the valuable work of the companies. It is for this reason that ultimately the effort to standardize failed. As a participant in the PMC effort explained:

> We had done this joint article with [two other personal genomics companies] and there was some suggestion of let's have some central registry where you would put your curated data once you gleaned the best odds ratio from the literature and put it in a central place where others can use it. But of course the problem there is if one company first does all the work then the other companies would be able to use that without putting as much effort into it. And that happened several times where we added some SNPs early on and we would see that [another company] added reports that brought theirs up to speed with ours. And its back and forth. So that kind of model may not really work very well.[76]

Companies devoted considerable resources to reading the scientific literature, aggregating findings from across studies, and evaluating their reliability. This work created the value that the companies offered: not raw genomic data, but *interpretations* of that data. They could not just give this value away and stay afloat. Recognizing these limits, the companies dropped their effort to standardize. Instead, they began to acknowledge, and even highlight, the subjective dimensions of evaluating and certifying studies.

Some members of the Congressional Oversight Committee noted this phenomenon with some amount of puzzlement in their voices. Why, they implicitly asked, did companies highlight their different standards? For companies, the answer was simple. Different strategies for evaluating data reflected their companies' interests, values, and intended users. Navigenics sought users interested in improving their health and set out to provide consumers with information relevant to health outcomes (what they describe as "utility and relevance") about which something could be done (what they call "actionability").[77] Thus, they conservatively picked SNPs that had strong associations with health outcomes and created algorithms that they could easily explain to their consumers. In contrast, 23andMe sought the digerati and those interested in participating in the new cutting-edge science of genomics. Thus, they used the same SNP discovery platform used by most researchers (the Illumina platform) and valued accuracy and transparency over utility and actionability. They were less conservative in how they chose SNPs and less concerned about the complexity of their algorithm. They assumed a consumer willing and eager to engage with the scientific literature, one who might even read published academic articles to interpret 23andMe's methods and results.[78]

Curation: Latest Fad or Long-Standing Practice of Science?

These different goals and values account for the differences in how 23andMe and Navigenics designed their algorithms and chose their SNPs. The term personal genomics companies used to describe this work is *curation*. The term is apt. Unlike objectivity, it presumes differences exist and that taste and values are needed to make sense of them. As a scientist employed as a *biocurator* explained to me: "I think most people have read some of the papers talking about the differences between the companies. What they take away is the companies give you different results and they don't always understand why. A lot of it is we have different criteria for what we put up on the website. I don't think that necessarily the general public or even scientists reading these have followed up and realize that it is different curatorial practices."[79] Unlike objectivity, which carries along with it the expectation of a single true answer, curations are supposed to differ—indeed, their difference constitutes their value. This collection of paintings, of records, of genetic variants has been carefully selected by a discerning eye, not the "God's eye" of aperspectival objectivity.[80] This curation distinguishes the collection and makes it valuable. Explained one senior scientist at a personal genomics company: "I think of it like an art curator. So, you're the curator of a museum. You have space where you're going to put some art. And your job is to go out and decide what art should be put in. And there's a lot of C-R-A-P out there. And there's some really nice pieces, and there's some nice pieces that are actually forgeries, and your job is to go out there and bring things in that are going to be of interest to people and useful."[81] In 2009 the *New York Times* reported that curation had "become a fashionable code word." and that everything from craft fairs to food stalls to sneakers now are curated.[82] Personal genomics companies added genomes to this expanding list.

While the use of the term *curation* might be a sign of personal genomics companies' entanglement in the latest marketing fads, it also might better describe what the scientists employed by them actually do: sort through vast amounts of data with a discerning eye that is informed by specific tastes and values. And this is not a new kind of activity. Curation long has been part of the life sciences. Careful collections of biological specimens and data have been central to biology from its earliest days.[83] If truth in advertising is what so many want from this new industry, perhaps in this instance, as uncomfortable as it might be, this is what they are delivering.

However, this story of the on-the-ground day in and day out interpretive work of companies remained largely out of the public view. Instead, reporters focused on discrepancies in results they received from different companies. Reportedly, they did not follow up to understand why the differences existed. As a biocurator at one of the companies put it: "You get papers published in *Nature*: 'Oh, look how different the results are!' And it's like yeah, well, you know they're different SNPs! And those papers were published without anyone calling any of the scientists at the companies to learn more. And I'm sure they were like 'Oh, well we don't want to be biased.' But, no, you're being naive. You're not doing a good scientific job by understanding all the variables."[84] Rather than a story about this fundamental problem of knowledge production, the story of personal genomics became one of the corporate corruption of science. Members of the US Congress and social scientist observers accused personal genomics companies of selling snake oil and mining for gold.[85] Scientists working for the companies, most of whom possessed degrees from prominent genome research centers at elite universities such as Stanford and UC Berkeley, reported that their academic colleagues also viewed the companies with suspicion.[86] Indeed, some who worked at Navigenics and 23andMe reported using their academic e-mails instead of their company e-mails when writing to their academic colleagues.[87]

At the Limits of the Corporate Corruption Story

Rather than delve into the ups and downs of different methods of deciding which genomic data to use (for example, data produced by the US National Cancer Institute or the Icelandic Health Database), which method for studying that data to trust (for example, GWAS or candidate gene analysis), and what criteria studies must meet to count as significant findings (for example, numbers of cases and controls, *P* values, and such), the debate focused almost entirely on who could be trusted to do genomics research. The verdict for most was clear: companies could not be trusted. Motivated by profit, they were the "snake oil salesmen" of the "high tech community" and posed serious threats to consumers.[88]

This representation proved a frustration for many in the companies who spent their days buried in the scientific literature, attempting to do what almost all in the genomics community believed needed to be done: interpret deluging genomic data. Some pushed back: "When you go to give a talk, if you're from a company, you have to list your disclosures. And then you have the academic person who gets up and says, 'I have no financial disclosures.' Yes you do! You get grants. In order to get tenure, you have to write papers."[89] Researchers who worked at the companies expressed frustrations with the black-and-white way in which their academic critics viewed the corporate world, and the assumption that freedom, and thus knowledge, only could be found in academia.

Yet critiques of corporate approaches to science highlight important differences and problems.[90] Companies almost always have a shorter temporal existence than public institutions. Of the three main personal genomics companies that led the way in 2007, only 23andMe remains open. And even 23andMe, despite its backing by Google, faces tremendous challenges. Citing failure to comply with FDA regulations as well as the "potential health consequences from false positive or false negative assessments for high risk indications" and failure of patients to "adequately" understand test results, on November 22, 2013, the FDA ordered 23andMe to "immediately discontinue marketing" its personal genome service.[91] 23andMe did negotiate an agreement with the FDA to continue to offer consumers access to their raw genomic data and genetic ancestry information but removed all health reports from the service. 23andMe vowed to find "the right regulatory path" for its consumers, and in June 2014 it submitted its first health report for FDA approval. In February 2015, it received FDA approval for its first genetic test, a test for a rare disease known as Bloom's syndrome.[92] Soon after, in October, it announced that it successfully raised $115 million in Series E financing.[93] At the moment, the company appears to be rebounding; however, there is of course a significant risk that it will not survive, and that all of its research and access to its data also will come to a sudden halt.[94] As a bioinformatician who had worked in industry explained to me: "In industry you can be on a project and for whatever reason you can get axed and all your work just gets lost."[95]

Companies also must at some point turn a profit. This is the corporate bottom line. 23andMe executives and board members have not shied away from this point, freely acknowledging to their customers that they are a "profit-seeking" company.[96] However, they believe that making money and making revolutionary science are not at odds. Yet at both 23andMe and Navigenics the need to create a profitable product did lead to cutbacks in the scientific staff. Of the six scientists I interviewed who worked at these companies, only one remains on staff. Most cited a shift in focus away from the scientific work of curation: "We had a shift in what we were focusing on in our products, so we didn't need as much curation. If the goal is to annotate the genome, then you would want to have a large staff of curators. If the company's goal is to survive, and to emphasize certain goals over others, then you may not share the same goal as being as complete in our annotation as [publicly funded efforts to annotate the genome]. There was a resource issue there."[97] While government-funded research also operates on bounded time scales, it is not as subject to the ups and downs of the market economy, and thus provides a more stable—but perhaps less revolutionary—approach to scientific research.[98]

A Radical Democratic Open or a Corporate Enclosure?

Could genomics benefit from both approaches? Many hope this will be the case, as it had been in other areas of biotechnology. However, others fear a less harmonious outcome. Instead of symbiosis, they foresee colonization. Certainly, statements such as this one from a staff scientist at a personal genomics company created worries: "I know nothing about [X] disease, but . . . generally our philosophy is gather a lot of data and try and let the data speak for itself, and find out what stories there are inside." For this scientist, lack of prior knowledge of disease X is not a problem, but a strength he celebrated: "I go around to give talks in front of biologists and say, 'Yeah, I know no biology. I'm proud of it. It makes my work better.'"[99] Assuming no prior knowledge, he argued, also creates greater respect for research participants through prompting researchers to look at *all* their data and recognizing that "they've got a history, and they're not just disease X."[100]

While claiming to respect the full complexity of their customers, many scientists, doctors, and indeed regulators felt personal genetics companies demonstrated little respect for them as persons, and for the repositories of scientific and governmental expertise—built up over decades—that they represented.[101] Like its major funder Google, many feared that 23andMe sought not to add to the efforts to make sense of human biology, but rather to take over. A January 2008 headline in the UK broadsheet *The Telegraph* captured this sentiment: "Google Wife Targets World DNA Domination."[102]

These concerns run deep, and move beyond 23andMe. Since the Human Genome Project days, biologists have worried about a Big Science takeover in which machines, money, and a mindless technical bureaucracy replace small-scale community-directed knowledge production.[103] Over the course of the HGP, a genomic bureaucracy grew up at NIH. Indeed, 23andMe attempted to mobilize the existing dissatisfaction with this established and powerful order to foment its research revolution. Rather than research that served the interests of this established order, Wojcicki and Avey promised research "of, for, and by the people."[104] While a compelling vision, many life scientists expressed concerns. Might this science for the people enact their very displacement and the takeover by machines and their multinational corporations?

Their fears had precedents in social thought forged three decades earlier. Lyotard in his *The Postmodern Condition: A Report on Knowledge*, precisely predicted that information-processing machines would forever change the nature of knowledge. Most strikingly for the story of personal genomics, Lyotard predicted that belief in state-supported learning would decline as the idea grew that societies only progress through circulating messages that are "rich in information and easy to decode." In this new world, the state represented noise in the system; it impeded "communicational transparency."[105]

Three decades later personal genomics companies argued that this kind of informational transparency should displace state-centered approaches to genomics. Easily accessible information, not consistent test results, the *Personalized Medicine Coalition* report on standards argued, should be the goal.[106] The state, and its efforts to enforce privacy and impede information flow, Wojcicki and Avey contended, was unjust.

Personal genomics companies were the first to put forward this vision of genomic information out in the open, freed from unjust state regulation. However, they would not be the last. Today, many of those in the Euro-American West—including myself—are being asked to take part in the breaking down of old regulatory regimes in order to create new unimpeded flows of data that our health providers promise will revolutionize medicine. It is to this current and final venture of the postgenomic condition that I now turn.

7 THE GENOMIC OPEN 2.0

The Public v. The Public

The trouble is that factual truth, like all other truth, peremptorily claims to be acknowledged and precludes debate, and debate constitutes the very essence of political life.
—Hannah Arendt, *Between Past and Future*

I had spent a good part of the previous decade researching and writing this book, talking to others about how to gather and analyze DNA such that the human genome might become meaningful and beneficial to all, and now the decision lay before me. In early 2013, I went to the University of California, San Francisco (UCSF) Medical Center at Mount Zion for my routine annual checkup. Eight months prior, then chancellor of UCSF, Susan Desmond-Hellman, penned the lead editorial in *Science Translational Medicine* in which she announced that patients were in the "best position" to understand the problems with patient privacy regulations and to demand the sharing of their data needed to "accelerate the cures all are awaiting."[1] She cited HIV/AIDS patient advocates as a notable example. At my doctor's office in January 2013, I had the opportunity to become another.

While checking in for the appointment, I was handed a clipboard, a pen, and the UCSF Terms and Conditions of Service form. With a bit of time on my hands, I decided to read the document rather than just sign it. I was half reading and half listening to the Cooking Channel tell me how to make the perfect polenta when I reached point four: "I also understand that my medical information and tissues, fluids, cells, and other specimens (collectively, "Specimens") that UCSF may collect during the course of my treatment and care may be used and shared with researchers." This is the sharing that Desmond-Hellman envisioned. It was, on her account, my chance to help my doctor make good on the promise of genomics to generate medical breakthroughs.[2] Yet rather than

UCSF Mission Bay campus (photo by Lindsey Dillon).

feeling empowered, I felt uncomfortable. Instead of signing the form, I stuffed it in my backpack and hoped I would still receive care.

What would it mean to have "all my medical information and tissues, fluids, cells, and other specimens . . . used and shared by researchers"? I lived on Potrero Hill and had watched the massive Mission Bay UCSF biomedical campus grow up over the hill in the former railyards of San Francisco.

I knew that the Bay Area ranked first in the nation for the number of biotechnology patents per million citizens.[3] I understood that Silicon Valley hoped that the next boom would come from harnessing infotech to biotech. I realized that many viewed collecting data and tissues from millions and openly sharing as a necessary and critical step. What I did not know was what all this would mean for myself and my fellow citizens. Despite being told by the form that I understood, I did not.

In the weeks that followed I discussed the experience with friends and colleagues. Many expressed surprise that UCSF stipulated sharing of personal data and tissue as a terms of service. Some encouraged me to write an editorial myself. I resisted. I did not welcome the prospect of publicly critiquing the policy of my employer. More than that, I feared not being the civic-minded, justice-seeking advocate of sharing that Desmond-Hellman imagined me to be. For many, sharing information and knowledge is at the leading edge of the twenty-first-century fight for a just, democratic public sphere.[4] Aaron Swartz, a chief architect of the Internet Archive's Open Library project and prominent member of the open source/open data movement, had died just a few weeks before, the victim some would say of the copyright wars.[5] Solidarity around the commitment to share felt particularly important.

Yet I did not feel that all injunctions to share arose from the same radical spirit. Would the causes of universal knowledge, justice, and health be served by signing the UCSF Terms of Service form? A space for public discussion of this question, I felt, needed to open. I decide to write.

On March 3, 2013, the resulting article titled "Should Patients Understand They Are Research Subjects?" appeared on the front page of the *San Francisco Chronicle* Sunday magazine, *Insight*.[6] Immediately I heard from strangers and neighbors alike. One person informed me that her refusal to sign a similar form led to a struggle with her medical provider over provision of care. Another wrote a letter to the editor, linking the question of whether UCSF should have the right to collect data and tissue from me to the question of whether Johns Hopkins should have had the right to collect cells from Henrietta Lacks. At issue, this *Chronicle* reader made clear, are questions about not just who benefits, but who profits.[7]

I personally was not concerned about who might profit from the collection of my data and cells. Indeed, it seemed unlikely that question would ever arise. However, I did worry about what was happening to medical care as it became ever more tightly harnessed to biomedical research.[8] I cared about what was happening to scientific research as it adapted to the new world of big data and its grand promises of medical breakthroughs.[9] I feared the erosion of the right to control—to own—one's body. Finally, I questioned what genomics' norm of openness meant when invoked not just to guide the action of scientists, but all people. The human genome should be in the open domain, but should *my* genome? The genomes of all people? What would be the value and meaning of openly sharing my tissues and data? Was it sharing or taking?[10]

This chapter seeks to answer these questions through exploring what happened when a group of citizens in the United States did say "yes" to sharing their DNA, medical records, and much more with researchers. These citizens took part in the Personal Genome Project (PGP), a Harvard University–based initiative led by George Church that aspired to sequence and openly share the entire genome of millions of citizens.

While the decade began with powerful calls to save the human genome from private enclosure, it ended with this effort to make public millions of human genomes. Unlike the HGP, however, the PGP did not garner widespread support. While publicly funded scientists could argue that it was unjust for corporations to use property rights to enclose the human genome, when scientists who received considerable private funding, like Church and Desmond-Hellman, argued that it was unjust for governments to use privacy rights to protect the free circulation genomic information, it produced less moral clarity.

The PGP found itself at the leading edge of efforts to rework the norm of openness amid these postgenomic conditions in which no clear line could be drawn between the public and private, good and bad, and heroes and villains. PGP proponents argued that privacy was hard to achieve and impossible to guarantee, and that just formulations of ethics flowed from that truth. However, truth and justice proved much more fraught affairs. What rights could citizens expect to enjoy in a world where the promised public goods of precision medicine required their personal information and their own flesh and blood? An answer to that question would not emerge easily.

As these questions and issues first presented themselves, I found myself at Duke University in the Institute for Genome Sciences and Policy. Robert Cook-Deegan, policy adviser to James Watson during the Human

Genome Project era, had just launched a Center for the Study of Public Genomics. Misha Angrist, who would go on to be the fourth person to have his entire genome sequenced and made public via the PGP (or PGP-4), had just arrived at Duke. It is in this milieu that I first learned of the possibility that my colleagues and friends might one day very soon have their genomes sequenced. I learned along with them about these brave new worlds being forged.[11] They would go on to write their own stories.[12] In the spirit of debate and dialogue needed to foster the public genomics that Cook-Deegan, Angrist, and others envisioned, below I offer my own.

From Bermuda to Bohemia: Openness 2.0

On April 25, 2009, the anniversary day of the Watson and Crick discovery of DNA, Harvard University launched a project that offered to sequence any resident of the United States over the age of twenty-one willing to openly share genetic, medical, and life experience data, and able to pass a test proving they knew what this meant. The project—dubbed the Personal Genome Project—sought to realize the long-standing dream of Harvard professor, and early genome sequencing guru, George Church: Genomes for All.[13] Church, who built an electronic computer during the 1960s at the age of nine, believed that everyone should be able to program the future. In the wake of the HGP, this meant programming not just computer code, but genetic code. It required not just a laptop computer, such as a MacBook Pro, but also a laptop sequencing machine, such as an Ion Torrent.[14]

To realize this vision, Church foresaw a radical reworking of genomics' norm of openness. While the final sequence of the human genome did end up in the open domain, the means of its production—the sequencing machines—remained firmly locked down in the Applied Biosystems Inc. monopoly. Church passionately argued that this ABI monopoly stifled innovation *and* democracy. Only a few centers could afford its expensive machines.[15] This concentration of power portended a dystopic social future, and it slowed the improvement of sequencing technologies. Those who sought to tinker with the ABI machines and innovate, Church reported, received threatening letters.[16]

Church set out to overturn this monopoly. He predicted that demand for sequencing would not dry up with the completion of the HGP, but would instead take off as all people took an interest in their genomes,

In 2012, Life Technologies developed an advertising campaign designed to demonstrate that its Ion Torrent was so simple and accessible that it could travel with you in a Mini Cooper (photograph used courtesy of Thermo Fisher Scientific; copying prohibited).

creating "a market of between one and six billion people with six billion base pairs." This would create more than enough demand to fuel competition and drive innovation.[17] However, there were two significant hurdles: the Clinical Laboratory Improvement Amendments of 1988 (CLIA) and the Health Insurance Portability and Accountability Act of 1996 (HIPAA). The first, CLIA, is a US federal regulation that governs clinical laboratory testing of human specimens and is intended to ensure the accuracy, timeliness, and reliability of laboratory test results. The second hurdle, HIPAA, is a US federal regulation that governs the use and disclosure of what the law describes as protected health information. Both impeded genome sequencers from giving genome sequence data to individuals, as most sequencing labs do not have CLIA certification or HIPAA authorization.

The Personal Genome Project sought to change this protectionist approach to genomic data. "This whole project," Jason Bobe, the director of the Personal Genome Project, explained to me in 2010, "is an engineering project around that problem. That's where it started."[18] Indeed, the first paper published by the PGP was not a traditional journal article describing research goals and results, but a policy paper outlining a new ethical approach to questions of genetic privacy. In a Perspectives essay published in *Nature Reviews Genetics*, titled "From Genetic Privacy to Open Consent," bioethics scholars Jeantine Lunshof and Ruth Chadwick and lawyer Daniel Vorhaus along with George Church argued that researchers no longer could deliver on promises to keep research subject data private. An open consent policy in which research "volunteers consent to unrestricted re-disclosure" of both their health records and their "genotype-phenotype data" was both more realistic and honest.[19] This policy, Bobe argued, was "the bedrock of the PGP."[20]

Open Consent, Open Technology

Open consent, PGP leaders imagined, would foster not just citizen science but also innovation. Indeed, Bobe explained that it was central to tackling "the next level of technological engineering problems."[21] To understand why, consider that when the HGP ended, a central source of DNA designated and consented for sequencing dried up. Instead of DNA in need of sequencing machines, there were sequencing machines in need of DNA. Without DNA to sequence because of privacy laws, there was no way to test and improve the machines. The PGP sought "to social engineer around this [privacy] problem" through making it possible for individuals to donate their DNA in full awareness that their privacy could not be protected.[22] Church, the first participant in the Personal Genome Project, recounts in his 2012 book *Regenesis* this link between the collection of his DNA for the PGP and the development of sequencing machines: "My tissue samples were taken in 2005 and 2006. My lab has developed or advised most of the current thirty-six commercial next generation sequencing technologies, and we test the technologies as they mature. The first set of samples was sequenced at Complete Genomics in Mountain View, California."[23] Complete Genomics went on to become a major player in the genome sequencing business and was bought in 2013 by the sequencing giant, the Beijing Genomics Institute (BGI). The PGP provided it with some of its first genomes to sequence.

Not only would PGP participants provide samples that would be used to test new sequencing techniques, they also would be used to test platforms for consuming the data. Church envisioned developing platforms in which information flowed freely between people and machines. PGP participants would receive free access to *all* their personal genome data. Indeed, through these platforms *all people* would have free access to all of the PGP participants' data.

For Church and other PGPers, no contradiction of logic or intent lay in this commitment to a *public* personal genomics in which companies played a central role. Times had changed since the days of the Human Genome Project. Companies no longer threatened to steal from citizens their common heritage through locking up the genome in property rights. Instead, they were important partners in an innovation process that gave as it took.[24] To give information of value to people, innovators first needed people to give of their DNA. They would use this DNA to test and develop faster, cheaper, and more accurate sequencing machines. The goal: a $1,000 genome that put genomics within reach of all. As Church argued in his 2006 "Genomes for All" essay: "The 'one thousand dollar genome' has become shorthand for the promise of DNA-sequencing capability made so affordable that individuals might think the once-in-a-lifetime expenditure to have a full personal genome sequence read to a disk for doctors to reference is worthwhile. Cheap sequencing technology will also make that information more meaningful by multiplying the number of researchers able to study genomes and the number of genomes they can compare to understand variations among individuals in both sickness and health."[25] Open access to DNA would lead to better, cheaper machines and ultimately genomic data that was meaningful and beneficial to all.

Church and his followers believed passionately in this open, inclusive approach, one that extended not just to the data but also to the machines. While in his early writings, including "Genomes for All," Church distinguishes between "individuals" and "researchers," ultimately, he and his followers foresaw a world in which all individuals could be researchers and access not just their data but the means of its production. As Bobe explained to me: "I imagine a future where the DNA sequencing technology and other technologies are totally decentralized. People access them and generate them themselves."[26] This was the radical world of open technology and Do-It-Yourself Biology (DIYBio) envisioned by supporters of the PGP.[27] Just as computers changed from mainframes owned and controlled by large institutions into personal computers owned and operated by citizens, Bobe predicted that genomics would transform from a massive government- and corporate-controlled endeavor to an everyday activity of citizens. It would emerge from genomics in a manner similar to the way that personal computing emerged from mainframe computing.[28] Eventually genome sequencing machines would be so cheap that we all would have them sitting on our proverbial laps.

For this to happen, however, engineers needed to break the medical and corporate monopoly on genome sequencing. This, PGPers believed, was imminently possible. As the Personal Genome Project's fourth participant (PGP-4), Misha Angrist, explained to me:

> MA: You don't need a multinational corporation to build a potato gun.
> JR: Are you saying a sequencer is a potato gun?
> MA: That's exactly what I'm saying. . . . I'm saying that biology . . . is a cottage industry. It's not Microsoft or Google or IBM.[29]

Boys used to launch potatoes in their backyards learning Newton's laws of motion, dreaming of outer space. Now they would create sequencing machines and learn the laws of genetics, dreaming of inner space.[30] This, Bobe asserted, "is a future that we will live in in our lifetime."[31] But only if we tore down the entrenched ways of dominant biomedical institutions that kept genome sequence locked up in the scientific bastille (for example, academic and medical research labs).[32]

This was a vision of a radically open science made possible by the fall of out-of-touch government regulations and the rise of cheap technology. In the future, unjust representatives (for example, ABI, NIH) would no longer regulate access to genomic information. All interested parties would have the power to look directly at their genome sequence using their own, cheap desktop sequencing machines located in community laboratories. As Bobe explained to me in the fall of 2010: "As a community, we're starting to build these community labs. There's another one I can't wait to visit, called GenSpace in Brooklyn. . . . They're just about to open their doors later this year and start selling memberships—like one hundred dollars a month and you can have access to the biolab. . . . They got an awesome space, in Brooklyn, in this real bohemian seven-story building that's got dancers and designers and cooks and . . . you can rent a room for cheap."[33]

Open Humans

But before there could be open science, first there needed to be open humans. In the world that Church imagined, these humans believed in creating an Internet of DNA that made the code of life readable to all.[34] As PGP-4 Misha Angrist explained: "Just as Linux code is available to everybody, I'm participant #1 and my code is going to be available to everyone. And he [Church] wanted to find other people who felt the same way."[35] Church believed that enough people—a billion or more—would be willing to create a web of information that knew no fleshly limits. This "data net"—brought into being by fast, cheap sequencing machines and open consent protocols—would produce an unprecedented growth in data that would exponentially increase the power of research.[36] In January 2014, the Knight Foundation and the Robert Wood Johnson Foundation awarded Jason Bobe and Madeline Ball of PeronalGenomes.org $500,000 to create this data architecture, what they and their collaborators called the Open Human Network. As Bobe explained to me in the summer of 2013: "[The] PGP will become one inter-operable study of open humans."[37]

Open Humans is similar in spirit and intent to Synapse, the data commons built by Sage Bionetworks, a nonprofit organization founded by former senior vice president of Merck, Stephen Friend.[38] The PGP gave John Wilbanks, former vice president for science at Creative Commons and chief commons officer at Sage Bionetworks, a copy of their open consent form. Wilbanks used it to craft what he dubbed Portable Legal Consent (PLC), a novel consent tool designed to change the agency of consent from what individuals let researchers do to their data to what individuals do with their data.[39] Rather than waiting for researchers to ask for it, PLC enabled people to initiate placing their data into a data commons. Similarly, Open Humans sought to let people control their data and share it with researchers. If enough people did this, Wilbanks predicted, the problem with private systems would become immaterial. The power of public domain data would overwhelm private enclosures: "If the infrastructure is in place for that [data] to get loaded into an open system and you've created the idea that patients should have a copy of their own data, then it doesn't matter about all this closed stuff that is happening. If enough people opt into a commons, then you've got competition in the open."[40] This, advocates of these new open data sharing platforms argued, would radically change how research gets done. Explain Bobe and Ball on the Open Humans website: "We think a critical mass of open humans will start a snowball of data sharing. Returning and publishing data can become standard practices that transform the public face of health research."[41]

The Open All the Way Down

While advocates saw open genomes and open humans as the second coming of the Internet, a force for human liberation and creativity, NIH and the Harvard IRB saw things differently. Reportedly, Jeff Schloss, who managed sequencing technology grants at the NIH, did not think that Church had adequately explained why the PGP should be allowed to move forward under different informed consent rules (that is, open consent).[42] Even more, when Google looked poised to provide the PGP with millions of dollars, NHGRI leadership reportedly told Google they thought the PGP would set personal genomics back by twenty years. Fully sequenced genomes simply had unproven value for individuals and posed too many privacy risks.[43]

The PGP's response to these concerns remained consistent: more openness. To make the risks as transparent as possible, it created a twenty-four-page single-spaced informed consent form. The form went into great detail—nearly six single-spaced pages—explaining the risks and possible consequences of personal data being widely used: allegations of nonpaternity, planting of one's DNA at a crime scene, denial of employment, contact by the press, and so on. The form also warned that all risks cannot be known in advance: "The Personal Genome Project is a new form of public genomics research and, as a result, it is impossible to accurately predict all of the possible risks and discomforts that you might experience as a result of your participation in this study."[44] Veracity, they believed, was of utmost importance.[45] They could not promise privacy because technically it was not possible. They could not promise full disclosure because the future was uncertain.

Nor could they place their trust entirely in informed consent forms. It was too easy to not read them and click "I consent."[46] Thus, to further ensure all PGP participants knew what they were signing up for, participants had to correctly answer *all* questions on the PGP entrance exam. The exam covered everything from the basics of genetics, to the history of human subject research ethics, to the specific practices and risks of the Personal Genome Project.[47]

In all these ways, architects of the PGP sought to build a project that upheld principles of openness and transparency.[48] Anything different, they argued, violated the liberal democratic commitment to truth. As Yaniv Erlich at the Whitehead Institute for Biomedical Research and

Latanya Sweeney of the Harvard Data Privacy Lab demonstrated, participants in genetic studies could be reidentified using only a few pieces of information, such as gender, zip code, and date of birth.[49] To promise privacy, the PGP leadership argued, was to lie to participants; it propagated ideology and illusions. The PGP sought to expose the truth. As Misha Angrist explained: "It [the PGP] is useful for pointing out that the emperor has no clothes, that absolute privacy and confidentiality are illusory."[50] In an age of Julian Assange, WikiLeaks, and Edward Snowden, PGPers believed that the NIH promised research subjects privacy at its peril.

> Things are going to break. Data is going to spill out, and the last thing you want to do is promise somebody [privacy] who, from their own goodness of their heart, because they've suffered cancer, or whatever, they want to give back and contribute to research. . . . If anything's going to set back genomics, it's not going to be the PGP it's going to be gaffes like that where there is suddenly national awareness that this project that millions of Americans are participating in is totally compromised and in it is the most intimate details of HIV, sexually transmitted diseases, predisposition to all this stuff.[51]

As the Ashley Madison hack would make painfully clear to millions of people in the summer of 2015, everything on the Internet could be out in the open, like it or not.[52] The future of genomics, PGP proponents argued, depended upon recognizing this truth.

From Truth to Practice

While clear in principle, how to create transparent and open infrastructures proved anything but straightforward. Even making visible information about what happened to PGP participant samples (for example, the date and times they arrived in laboratories for processing) posed significant challenges. It required the PGP's partners to design this transparency into their informatic systems, a task that required time and energy that was not always forthcoming. Coriell, for example, which transformed PGP samples into cell lines, only provided patchy information about PGP samples.[53] As in the case of the Bermuda principles, making data available required commitment of resources, and not everyone agreed when and where this made sense.

Differing values also structured what information could be made available about the sequencing machines. Church sought to create a platform to innovate what he called open technology. He believed that everything about the process—from designing the machines to consuming their outputs—should proceed in the open. But the companies who supported the development of the Polonator, Church's next-generation sequencing machine, balked. Church wanted to bring the Linux approach from software to hardware and keep the Polonator in the open domain. Agencourt Personal Genomics purchased a license to the patent from Harvard. At the time, they agreed to keep the Polonator open source. However, soon after Agencourt received the license they began to question the open-source approach. A few years later, in an attempt to maintain their position in the sequencing market, ABI—the very company whose monopoly on sequencing machines Church sought to break—decided to develop machines using the Polonator's polony (polymerase colony) approach and bought Agencourt. Agencourt immediately dropped the Harvard license. A member the Church lab explained the underlying dynamics: "Open technology seemed a little premature because open hardware is a little more difficult. Software is great because it is trivial to copy and trivial to edit and share your copies. . . . It costs something to buy hardware. There are the infrastructure costs . . . and then there are reagent costs. . . . And it is physically man-hours very difficult to put together. Software is the opposite."[54] Church dreamed of bringing the cost of sequencing machines down so that the cost of hardware neared that of software, making it accessible to all. However, the reality was that these machines were far from free. As in the HGP days, they continued to require tremendous investments of money and labor, and more than just a little bit of devotion.[55] As a result, despite Church's best efforts, they did not circulate in the so-called free economy.[56] They remained the products of Fortune 1000 companies—in this case Applera, the company that bought ABI.[57]

Coriell and ABI were not the only ones who limited the free flow of information. The PGP itself set limits as it faced resource constraints and reckoned with its own values and goals. Making data open required resources, and so decisions had to be made about whether, for example, participants could access all the raw data generated from sequencing their genome. As Bobe explained:

> It was a debate early on. It was purely a practical thing that the rawest computable data were actually terabytes and terabytes per genome of image data. . . . Even before that there are dots of light that some

> processer says "Oh, this spike is an A and this spike is a T." . . . So one of the people who is involved in the Church lab's effort came from astronomy and was a woman who specialized in math. [She] took dots of light which were stars and measured those dots of light in ways to predict their distance, their composition. . . . [The PGP] is looking into the other end of the telescope into the microscope of the Polonator. That same math applied. So she was able to come up and get a much better ability to pull data out of these images. . . . That was great, but it was too resource intensive and seemed like a waste of time to do that for everybody. So there are those decisions that get made.[58]

The PGP did not itself seek to improve sequencing machine algorithms but rather to create a platform in which others might. Thus the PGP leadership decided against storing and providing open access to the raw sequencing reads of PGP participants.

In addition to shaping what information existed in the open, values and project goals shaped which humans participated in the PGP's open domain. While Church, Bobe, and Angrist sought to open the doors of genomics up to anyone who could meet the open consent criteria, PGP participants today are far from anyone. When I asked Bobe in the summer of 2013 who made up the PGP, he responded: "It is largely white. It's not in parity with the population. So with the PGP10 we had an African American, and so that's one in ten in the population and we were in parity for a very short time! But there are very few African Americans that are involved. I don't know exactly what the percentages are of other ethnic groups, off the top of my head, but I know in particular African Americans are very rare." Indeed, Coriell classifies 85 percent of its PGP samples as "Caucasian," less than 1 percent as "African American," less than 1 percent as "Chinese," 2.5 percent as "Asian, Other," and less than 2 percent as Hispanic American.[59]

Despite ongoing concerns about the lack of diversity of biobank collections, for the PGP this lack of diversity was not a central concern.[60] As Bobe explained: "We haven't done any recruiting. Diversity is certainly important to research but I don't really want to be in a position where I'm trying to convince people to share data when they don't want to."[61] The freedom to act openly without inducement or constraint was the dominant value that structured the PGP. Diversity was important, but it could come later. For now, the PGP had more than enough people—at least ten times more than it could possibly sequence—to build out its system and to learn how to scale. "We are still just not in the position of wanting or needing to go recruit at this point. We are still building and thinking about scale and hopefully the problem of diversity can be solved in a couple of different ways. It's an issue, but it's not a burning issue for us yet."[62]

Open, but Inclusive?

The PGP grew out of a grand vision to build an open platform for genomic and biomedical information, one that would liberate technological innovation and restore the world of open science that Sulston feared lost. It promised a revolution that would overthrow the NIH regime that protected elite researchers and entrenched bureaucrats. It sought to create a science of, by, and for the people. Symbolic of this vision, on June 14, 2011, Jason Bobe and others central to the DIYBio movement hosted what they called the DIYBio Continental Congress of North America. Invoking the legacy of the American Revolution, "individuals and delegates" gathered to discuss basic principles.[63] In both this new polity of biology and that of the new American republic, founders believed citizens should possess a moral quality—or "virtue"—that made them able to set aside self-interest for the pursuit of the public good.[64] In practice, these citizens then and today also were either entirely or mostly white men. As a PGP participant explained at the 2014 PGP-organized Genes, Environment, and Traits (GET) conference, most got into the PGP because of their interest in "fancy science." They were technoscience enthusiasts, a group that today still falls far short of representing human diversity.[65]

As in the early days of the American republic, this lack of diversity raised questions about the PGP's implicit and explicit claims to justice.[66] At the GET 2012 conference, these questions lay barely beneath the surface. As I described the tension in my field notes, meeting organizers claimed to be breaking down the walls around genomics, letting in nonexperts. Yet despite this, the walls still seemed high. Looking around, the crowd largely consisted of a younger set dressed in crisp shirts and jeans and an older one sporting khakis with jackets. Even when Esther Dyson took the stage and announced that her panel would move beyond the in-group discussion to ask what the "rest of the world" was making of the PGP, it was still all men, and only one who appeared not white, who made up this "rest of the world." Organizers proclaimed the walls down. Panels such as these made them look still up.[67]

I would not be the only one to notice. In a breakout session titled "Now You've Got Your Genome, Now What?," a PGP scientist explained that PGP-10, a self-identified African American, "gets a lot of weird things" because he is an African American. Nothing more is said about this until I raise my hand and ask why. The PGP scientist responds that most genomics studies are done on "Caucasians," and so "African American" variants appear unique. A PGP participant in the audience responded: "This is a problem. What is the PGP's response going to be?" The PGP scientist explained, as Bobe had to me, that they wanted people to freely choose to take part in the PGP. Thus, they were not pursuing recruitment of anybody, including those underrepresented in the PGP.[68]

In its radical embrace of openness, in the eyes of many the PGP inherited the mantle of "public genomics" from the HGP.[69] It was led once again by a benevolent bearded scientist, this time not Sir John Sulston, dubbed a knight by the British crown, but George Church, hailed a god by the American comedian Stephen Colbert.[70] Yet just below the surface many felt there was something not quite right. Just as the project would not include all information, it would not include all people. Those exclusions made manifest the main values and goals that shaped the PGP: technological innovation; fast, cheap and widely accessible genome sequencing. While Church spoke eloquently about ethics, at heart, as many close to the project explained, his main goal was to advance sequencing technology: "Does he [Church] have lofty philosophical ideas in his head? He absolutely does but at heart he is an inventor. He's a technologist. He's a tinkerer. . . . The PGP was going to be the instantiation of his sequencing technology."[71] Bobe concurred: "The PGP was started because George was a technologist and he wants to evaluate new genomic technologies against human samples and see how well they do."[72] It did not matter which human samples. It mattered how many and how fast. As a 2010 PGP publication explained: "The PGP aims to produce public genomics research—and to develop and evaluate associated technologies and research—on a large and expanding scale."[73]

The Public v. the Public: Contestation over the Constitution of Science and Justice in the Postgenomic Era

While for technology development the question of "which humans" did not matter, for science and justice it did. A mostly white participant base returned the PGP to the problems genome scientists faced at the beginning of the decade: mostly white and male scientists and engineers built the field of genomics, and the DNA in its databases reflected this. The diversity of the genomes stored in genomic databases affect scientists' abilities to interpret them.[74] Much like the problem Georgia Dunston faced with HLA typing in the 1980s, lack of data on African American bodies led to difficulties knowing, and thus caring, for African Americans.[75] Building a public database in which most participants are identified as "Caucasian" might exacerbate this existing disparity in health information. While the PGP might be open science, would it be a just science? Whose interests would it serve?

In claiming the mantle of public genomics, the PGP clearly signaled that it intended to serve the public. The description of the HGP on the Harvard website makes that message clear: "We believe sharing is good for science and society. Our project is dedicated to creating public resources that everyone can access."[76] Harkening back to the Mertonian norms of science, Church and the PGP embraced the norm of openness. However, unlike Merton, they did not see technology at odds with this norm. Indeed, they believed just the opposite: technology could foster openness. This is not because they believed that the notion of technology as property enshrined in *U.S. v. American Bell Telephone Co.* no longer applied. They did believe that inventors could patent their inventions. Indeed, Church advised and received support from dozens of companies, including large pharmaceutical companies such as Merck and Eli Lily.[77] Since 1989, he had been named as in inventor on sixty-four patents.[78] He did not oppose patents and private property. To the contrary, he believed his patented technologies could foster the free flow of information—for example, by making the information contained in DNA easier to read.

Instead of corporations' *property* rights, Church and the PGP argued that the government's *privacy* rights impeded openness. In a reversal of the dominant story, they positioned the public sector, not the private one, as the impediment to delivering knowledge and power to the people. This time the battle was not between the public and the private, but between different efforts to forge genomics in the public interest.[79]

In an age of corporate-funded science, acting in public, and in the public interest, still powerfully defines good science.[80] However, what "the public" and "public interest" entails are a matter of much contestation. The outcomes of these contests are consequential. Indeed, some PGPers suspected that NIH argued their project threatened privacy rights—and, thus, the public interest—in order to not open the door to challenging established modes of doing research: "Why are we not funded? I think part of it is that it's just so nonstandard to do stuff like this. By funding it there's kind of an implicit statement that a lot of existing research is going ahead with the wrong format."[81] The PGP's invocation of "the public" similarly acted to promote their approach to genomics.

Whatever the veracity of these speculations, what is clear is that simple oppositional narrative frames—whether public v. private or public v. privacy—miss the on-the-ground complexities of contemporary struggles to forge knowledge and justice. Rather than oppose, at least in the Euro-American West, many philosophers, political theorists, and judges understand both property and privacy rights as fundamental to good public governance. Founders of the American polity, for example, believed that a private space free from public inspection was critical to cultivating independence and freedom.[82] Yet while a space of one's own proved of utmost importance, it was not the only value.[83] As the Supreme Court of Tennessee asserted in *Vaughan v. Phebe*: "Freedom in this country . . . is not confined in its operation to privacy." It moves "from the kitchen and the cotton field to the court-house and the election ground."[84] In short, freedom had private and public enactments.[85] This point did not prompt debate. How to properly delineate the private and the public sphere, and the nature of freedom in each, did. Indeed, arguments about the proper bounds and meanings of the public and the private have been at the heart of critical thought and debate in liberal democracies for centuries.[86]

What strikes many as radical about the PGP is its assertion that privacy is no longer something that the government, or any entity, can assure. This indeed was the clear message of a *New York* magazine article that Angrist sent my way in the summer of 2007 titled "Kids, the Internet, and the End of Privacy."[87] While *New York* represented their argument as only understandable to those under thirty, social theorists have long expressed anxieties about the implosion of the private and public realms.

Arendt dated this phenomenon to the rise in the "modern age" of "scientific thought," which enabled nations to conceive of government as an efficient administration of a "super-human family" called "society." Then the disruptive force was not social media, but a social economy.

> [The] emergence of the social realm, which is neither private nor public, strictly speaking, is a relatively new phenomenon whose origin

> coincided with the emergence of the modern age and which found its political form in the nation-state. . . . In our understanding, the dividing line is entirely blurred, because we see the body of peoples and political communities in the image of a family whose everyday affairs have to be taken care of by a gigantic, nation-wide administration of housekeeping. The scientific thought that corresponds to this development is no longer political science but "national economy" or "social economy."[88]

Arendt, a German Jew, wrote in response to the Nazi rise to power in the midst of the economic crises of the late 1920s and early 1930s. The unemployment rate in Germany soared to 33 percent. Millions worried about where their next meal would come from. The Nazi party promised a national economic solution. It made providing for the biological needs of families a central function of the state. Necessity, not freedom, moved to the heart of collective life. The result: ordinary citizens confronted the dilemma Hans Behnke faced in Wolfgang Staudte's critically acclaimed 1949 East German film, *Rotation*.[89] Behnke's refusal to join the Nazi party cost him a job he desperately needed to feed his wife and newborn son. He later gave in, joined the party, and once again could put food on his family's table. He could, however, no longer speak freely. Arendt sought to understand the causes and consequences of this loss of freedom of expression, and its replacement by a mass administrated society driven by fear and necessity. In this mass society, the distinction between the private and public realm is lost, and with it the freedoms guaranteed by each realm.[90]

Of course, contemporary concerns differ radically from those of Arendt and Staudte.[91] However, what is similar is that once again questions of life and death animate calls to give up the rights of a private realm for a greater public good. A visit to the doctor's office, for example, is not just a space in which one is cared for, it is a space in which one is called to care for others. I am asked to give my tissues and DNA to my medical provider in order that I might advance medical research and improve health care. If I do not give up my privacy rights for this public good, at stake is not food on my table, but my medical care.

The PGP might appear to avoid these dilemmas through clearly stating on its informed consent form that it does not exist to provide for the medical needs of participants. Indeed, the PGP consent form tells

Billboard in SOMA District of San Francisco, California, January 25, 2015 (photo by author).

participants that they "are not likely to benefit in any way as a result of [their] participation in the PGP."[92] Yet some leaders of the PGP recognize that they cannot be sure that all participants fully understand this. It is, they admit, likely that some join the PGP because of concerns about their health.[93] Others acknowledge that if efforts like the PGP succeed, it will be because of the support of rare disease communities, communities motivated to share their DNA and data because of hopes for new treatments.[94] Finally, the PGP operates in a broader milieu in which large biomedical institutions like UCSF explicitly tie support for public genomics to issues of life and death. In its MeForYou campaign, UCSF seeks to "rally the public" to create "a new social contract that calls on people to take ownership of their own health and the health of their loved ones."[95] This means sharing their bodily tissues and medical records so that the young girls like Georgia featured in their MeForYou campaign might live.

This is the broader context in which today we are called to share data. In the course of their everyday lives, citizens—at least in the United States, the United Kingdom, and Europe—may encounter claims that genomic data will improve health care and even lead to the cure to cancer. I, for example, came across this billboard as I cycled home from a movie in January 2015.

The billboard stood at a busy intersection in San Francisco that thousands of Bay Area residents pass by every day. Perhaps later while on

Facebook these same inhabitants of the Bay Area will find out how they can help the University of California's efforts through sharing their DNA and medical records. They will learn that not only can they help science, but also they can help their communities, and those they love. Me For You. *Gemeinnutz geht vor Eigennutz*.[96]

The good of the community is linked to the good of science in a manner that acts to compel consent. While I would not lose my job if I did not sign the Terms of Service form put before me at my UCSF doctor's office, I did worry about receiving medical care. I also felt the moral weight of the injunction to share. Did I not want to help my fellow citizens? Dissent, or even public dialogue, under these conditions becomes difficult.

End of Privacy, End of Story?

Lessening individual and collective investments in privacy today promises new expanded public domains. Yet what kind of public domains are these? Can they, for example, foster the value of public dialogue that long has been at the heart of democratic imaginaries of good public government?

Consider the "dis-ease" among PGP participants created by Harvard Privacy Lab's Latanya Sweeney's re-identification of several hundred PGP participants. In an online public forum hosted by Harvard Law School's Bill of Rights Project, Michelle N. Meyer, a member of the PGP Board of Directors and a participant in the study, raised questions about whether Sweeney's actions were ethical. While PGP participants agreed to put all of their genomic and personal data into the public domain, most also chose not to reveal their names. While they understood that there was a risk that they would be identified, Meyer argued that they did not consent to the intentional revelation of their identities. She explained:

> In this entire episode, what has surprised me the most is how readily so many people have concluded that assuming the risk of re-identification constitutes giving permission to be re-identified. Consider some of the other risks assumed by PGP participants. . . . Surely the fact that I acknowledge that it is possible that someone will use my DNA sequence to clone me (*not* currently illegal under federal law, by the way) does not mean that I have given permission to be cloned, that I have waived my right to object to being cloned, or that I should be expected to be blasé or even happy if and when I am cloned.[97]

PGP-4, Misha Angrist, a member of the online forum, responded: "Of course not. No one is asking you to be silent, blasé or happy about being cloned (your clone, however, tells me she is "totally psyched"). . . . But I don't think it's unfair to ask that you not be surprised that PGP participants were re-identified, given the very raison d'être of the PGP. And yet you were. As your friend and as a member of the PersonalGenomes.org Board of Directors, this troubles and saddens me."[98] To which, Meyer replied: "Here's the thing: some people *did* say that PGP participants have no right to complain about being re-identified (and, by logical extension, about any of the other risks we assumed, including the risk of being cloned)."[99] It was this attempt to shut down and silence that worried Meyer. She sought instead to create a space in which one could ask under what conditions re-identification might be ethical: "Exploring that question is the aim, in part, of my next post. What I tried to do in the first post was clear some brush and push back against the idea that under the PGP model—a model that I think we both would like to see expand—participants have given permission to be re-identified, 'end of [ethical] story.'"[100]

In their 2008 *Nature Reviews Genetics* article, Church, Vorhaus, Lunshof, and Chadwick laid out the central position of the PGP: veracity is a necessary condition of consent. Researchers must tell the truth, and that truth is this: "The data can be and likely will be accessed, shared and linked to other sets of information, and the full purpose and the extent of further usage cannot be foreseen."[101] There is no guarantee of privacy. Your data will be used in ways that cannot be predicted or controlled. Act in accordance with this truth and ethics and justice will follow.[102] End of story.

But, of course, there is always more to the story. As Meyer pointed out, there is a difference between understanding that others might use her data to do all manner of things and sanctioning others to do these things. Surely, she argued, the PGP commitment to transparency and truth should not lead to *less* control for participants: "It would be bizarre if the PGP and other data holders who follow suit were punished for their frankness by a policy that treats disclosure and acknowledgement of risk by data holders and data providers, respectively, as permission that obviates the need for consent."[103] Another world, she believed, was possible. And indeed, it was. In the fall of 2014, major funding bodies in the United Kingdom—including the Wellcome Trust—recognized the importance of telling research subjects that they could not guarantee privacy *and* took steps to deter re-identification.[104]

The PGP position that privacy is impossible to secure, and that principles of openness, not privacy, should guide governance of tissues and data, made it difficult to debate and consider these other possible worlds. While admirably motivated by the ideal of truth—and justice—these clear positions curtailed explorations of the many and varied enactments of privacy, and the ways in which these enactments impede *and* foster the creation of public goods. Privacy does not always oppose publicness. To the contrary, as the anthropologist of science Cori Hayden's decades-long study of privatization in the life sciences reveals, privatization has generated "a riot of collectivization."[105] So too has collectivization spurred new forms of privacy.[106] Why have these possibilities at times been hard to consider within the PGP?

Arendt's reflections on truth offer some insight. "The trouble," she writes, "is that factual truth, like all other truth, peremptorily claims to be acknowledged and precludes debate, and debate constitutes the very essence of political life."[107] In the case of the PGP, a commitment to the factual truth that data could and would be shared made it difficult to debate what this meant. Those who authored the PGP protocols believed PGP participants should be comfortable with this truth or they should withdraw.

This position left too little space to explore and debate the choices inevitably made as the PGP enacted its version of openness and public genomics. As in the case of the HGP, placing genomic information in the open domain required resources: if nothing else, it required server space and electricity. Not enough of these basic resources existed to store all of the information. PGP leaders made decisions about what to keep and what to discard. Similarly, not all uses of the information could be supported. Requiring participants to know and be comfortable with the fact that their data could be used for purposes the PGP could not foresee did not encourage debate about the ways in which the PGP acted to deter and foster uses of the data.[108]

Opening the Story: Open Science as Open Debate

While the norm of openness has a long and much-celebrated history, especially in the life sciences, its enactment first in the Human Genome Project and then in the Personal Genome Project raises fundamental questions about its meaning and ongoing value. Today openness is not a counternorm to technology but rather a norm that fosters the very technological innovation that is the beating heart of the life sciences.[109] For sequencing and other biotechnologies to advance, tissues and data must freely flow. Under these conditions, as the stories of the HGP and PGP demonstrate, the ethical and epistemic value of openness changes, raising critical questions: While genome scientists' enactments of openness might foster the advancement of machines, do they support "good science" and the norms of free inquiry?[110] Might the embrace of openness unwittingly install scale and efficiency as the driving imperatives of genomics, displacing the broader goal of building a genomics that is of, for, and by the people?

As the decade after the sequencing of the human genome began, belief in the tremendous meaning and value of genomics for all people held at bay exploration of these more critical questions. As the decade continued and it became clear that the postgenomic condition would be one in which the meaning of genomic data was far from clear, a space for critique opened up. Church and the Personal Genome Project powerfully occupied that space, arguing for a new approach to public genomics and openness. The NIH, they quipped, stood for Not Invented Here. Forget about privacy.[111] Long live The Open! These were big and bold ideas that inspired many and played a critical and important role in opening up the debate about the proper practice of genomics, technology development, and data generation and use. Much was and continues to be learned from the PGP's willingness to challenge long-standing beliefs about the proper ethical conduct of genomics research and data sharing. Yet ultimately the PGP's approach to truth and justice confronted limits.

How do we respond to and act beyond those limits? How might we better attend to the specific cuts and choices our enactments of openness make: to the practices of doing science supported and denied; to the knowledge gained and forsaken; to the machines and people who garner our care and attention, and those who fall further outside of our collective concern?[112] What might this look like, and how can it be achieved? It is to these questions that I now turn.

8

LIFE ON THIRD
Knowledge and Justice after the Genome

It is as though we have fallen under a fairyland spell which permits us to do the "impossible" on the condition that we lose the capacity of doing the possible, to achieve fantastically extraordinary feats on the condition of no longer being able to attend properly to our everyday needs.
—Hannah Arendt, *On Violence*

In the decade since the completion of the human genome sequence, the world has experienced dramatic and polarizing change. For most, these have been times of crises. Turn-of-the-millennium optimism about global prosperity quickly transformed into economic and political shocks.[1] The financial crisis of 2008 ushered in austerity programs in institutions in every sector, including my own. Employers slashed pay. They asked workers—whether white or blue collar—to work much more for much less. Millions lost their homes. Millions more, their countries.

For the few, these have been times of great prosperity. Today it is hard to imagine a world without Facebook and Twitter. However, a mere decade ago these companies did not exist. Neither did the countless start-up companies spawned by the social media industry. The very few, and the very young, became billionaires. Thousands more became overnight millionaires.[2]

Genomics is a product of these times. It grows out of and embodies this disjunction between great riches and deep deprivation. Perhaps nowhere is this more evident than where I have lived this last decade, the lands south of Mission Bay in San Francisco. Where once there were abandoned train and car lots and oil drum fires, where only a decade ago it was just about possible for a young University of California (UC) assistant professor to afford a small apartment, there now are gleaming buildings that exude hope, promise, and prosperity. In Mission Bay, former home of San Francisco's working port and train yards, the city in conjunction with University of California, San

Third Street Building Scape (photo by author).

Francisco (UCSF) and venture capitalists built 3.1 million square feet of labs and offices, a 289-bed UCSF Benioff Children's Hospital and 1.7 million square feet of biotech space.[3]

The sequencing giant, Illumina, commands a building the length a city block. The world's largest drug maker, Pfizer, occupies 11,000 square feet, and Bayer 50,000 square feet more. To support this growth, the city of San Francisco invested $400 million in roads and infrastructure and installed a $648 million Metro Line on Third Street to provide transit to Mission Bay.[4]

Yet just a mile farther south down Third Street, the metro line may run, but intense investment ends. Here in Bayview–Hunters Point the socioeconomic and health indicators are some of the worst in the nation. A report by the Brookings Institution in 2011 described the Bayview as an area of "extreme poverty" where 40 percent live below the poverty level.[5] It is also an area with high cancer, asthma and morbidity rates. The naval shipyard—which for decades employed thousands—served as the home base for the ships that carried the atomic bombs to Hiroshima and Nagasaki. Radioactive waste from sandblasting these ships, as well as toxic materials from the city's power plant, sewer, dump, and numerous other industrial facilities pollute the land and water.

Asthma hospitalization rates are four times the state average. Infant mortality rates are the highest in all of California. Breast and cervical cancer rates are twice those in the rest of the Bay Area.[6] The contrast, which I experienced many times while riding my bike down Third Street over this last decade, is shocking. How is it possible to invest so much in in-

Homeless encampment, UCSF "Bay Campus" (photo by author).

novative science and technology and so little in the people who live in its shadows?

This is the question raised for many in the wake of UCSF's announcement on March 1, 2016, that it planned to close the New Generation Health Center. The center serves people in the Bayview–Hunters Point neighborhood, and is the last full-service reproductive health clinic serving poor black and Latino youth in San Francisco. The university cited "financial reasons." However, as Joi Jackson-Morgan, the deputy director of at the Third Street Youth Center and Clinic, argued: "They said it costs too much but they can put this new facility in Mission Bay?"[7] Or as a petition posted to ColorOfChange.org put it: "The message the city of San Francisco and the University of California San Francisco are sending is blatant: money matters, not Black and Brown lives."

While the power and promise of genomics and precision medicine gleam brightly between China Basin and Cesar Chavez Streets on the Third Street light rail line, what they will mean for those who live in the blocks, regions, and nations beyond is far from clear. What the stories I have told in this book do make clear is that from Tuskegee, Alabama, to Glasgow, Scotland, to Ibadan, Nigeria, the people who gathered together in town halls, community engagements, and online to discuss genomics as well as genome scientists cared about who these brave new worlds of science and technology would benefit and what its effects

Abandoned shipyard buildings, looking back on Illumina building (photo by Lindsey Dillon).

would be in people's everyday lives. Some worried genomics offered no benefit, or that it would do harm. Might biotech development raise land values in Mission Bay, forcing out longtime residents, and the community organizations, like New Generation, that served them? Might building temperature-controlled environments for data matter more than maintaining and recuperating environments for humans and nonhumans alike?[8] Others worked to make genomics research beneficial in the present through negotiating support for community health care in exchange for their blood and DNA.

The meaning of efforts to listen to citizens' voices also remained unclear. For some, the NIH's commitment of millions of dollars and thousands of staff hours to develop community engagements created hope for a genomics that espoused and supported democratic ideals. The commitment of Generation Scotland scientists to engage citizens created similar goodwill and optimism, at least among some of the social scientists with whom they worked. Others expressed skepticism from the start. "What does the government want with us this time?" asked some involved in the Communities of Color and Genetics Policy Project in the United States.[9]

In all cases, whether begun with hope or resistance, efforts to democratize genomics encountered more fundamental concerns. Scientists, policy makers, and citizens all took first steps down a path that they hoped would create a genomics that was of, for, and by the people, but quickly encountered problems of justice and knowledge that made the onward route unclear. Tuskegee researchers did open the doors to including African American people and their genomes in genomic research, but ultimately the samples were not used and the effort abandoned. Generation Scotland avoided the international outrage sparked by efforts to study the genomes of the people of Iceland, but only to create a biobank with unclear value for both the people of Scotland and researchers. 23andMe initially successfully positioned itself as a democratic maverick, but today faces a class action lawsuit.[10]

Rather than a story of progress in which genomics and democracy come together to humanize and advance genomics, today we live in the midst of changes that are creating a new order that the language of democracy helps to facilitate but fails to describe. How, as Arendt asked, can we now think and speak that which we do? How can we articulate and address the fundamental questions of justice, ethics, and knowledge that lie before us? *The Postgenomic Condition* offers some insights.

From Information to In-Formation

We live in times where attention to poetics may matter more than principles. Poetics derives from the Greek word for "to make," *poiesis*. It turns our attention from a world already made, whose general principles are known, to a world in the making, where meaning, let alone principles, are far from secured. It invites us not to assume meanings, but to creatively play with words and worlds. The stories in this book offer a postgenomic poetics that might help foster this collective creation. In this spirit, I play throughout with the suggestion that it may be more illuminating to think of this time after the human genome as not part of the information age, but as an age "in-formation" where little is known, and much is promised. It is a time of excitement generated by unmooring past certainties and anticipating radical new futures.

Yet in the name of these futures, much also is undone. As the political theorist Wendy Brown observes in her latest book, *Undoing the Demos: Neoliberalism's Stealth Revolution*, we live in times where democracy is too often divorced from politics—that is, it is separated from the struggle over the fundamental goals and principles of a society.[11] The longstanding goods and principles of liberal democracies—inclusion, participation, freedom—have become procedural and transactional, cut off from the powers and practices needed to make meaningful change.

Genomics, I suggest, is a part of, and not apart from, these fundamental transformations. The field's understandings and enactments of information are emblematic. Genomics promises to deepen democratic powers of self-control through creating vast new stores of life-enhancing information about ourselves. This is in marked contrast to today's on-the-ground experiences with genomic data. As Misha Angrist, PGP participant number four, discovered, a whole genome sequence was too big to open on a personal computer.[12] To even "see," let alone "know" the data requires access to universities or companies that have enough computing power. Even with the support of Google, the power of interpretation largely eludes even 23andMe. The reason, as the stories in this book make clear, is that genome scientists built infrastructures designed to encode the sequence of our DNA into 1s and 0s on silicon, not to decode its meaning. While in the decade after the sequencing of the human genome many sought to shift genomics' focus to interpretation, the money still primarily flowed to those who could produce genomic information more quickly, cheaply, and efficiently. Informatic capitalism took a powerful hold. Thinking about meaning and ultimate purposes lost legitimacy—a frivolous activity that threatened to waste time, and thus lives.[13]

Privatization of the Public Realm

Genomic infrastructures also subjected the concept of the public to a similar transformation, unhinging it from its liberal democratic meanings and harnessing it to the instrumental ends of informatic capitalism. In the political thought of the Euro-American West, as Arendt demonstrated, the word "public" long has denoted that which "we" all can see, and thus that which we can all gather around to build a common world.[14] "It is the publicity of the public realm," she wrote, "which can absorb and make shine through the centuries whatever [*sic*] men may want to save from the natural ruin of time."[15]

As in past times, in these times world leaders argue that making things visible is inseparable from building a meaningful, governable, enduring common world.[16] On his first day in office, US President Barack Obama pledged that he would hold himself to "a new standard of openness." Transparency, along with the rule of law, he told the American public, would be one of "the touchstones of this presidency."[17] Obama defined digitization of medical records as central to this goal of creating an open style of public government. His administration invested $30 billion alone in encouraging hospitals to digitize their medical records to create "meaningful use" of these records.[18] The idea: increasing the visibility and circulation of information through digitization leads to good government and better medicine.[19]

Yet today it is no longer clear that the link Arendt observed between publicity—or making visible—and the creation of a public realm—a realm made up of things with enduring, common value—still holds.[20] Does making patients' medical information more transparent and free-flowing lead to the creation of things of value for the American public? Sometimes the answer to this question is "yes."[21] For example, making public different breast cancer genetic variants found through genetic testing would help more women know if they are at a high risk of this deadly disease. However, more generally, the benefits of making medical information more available through digitization are less clear. In 2005 the RAND corporation predicted that digitizing health records would save the United States $81 billion annually. In 2014 it found that digitization led to no savings, and instead, might increase costs by making it easier for doctors to order tests.[22] Further, while making information more visible might lead to more information, how doctors and patients alike should sort through this information to create good medical decisions is far from clear. How one, anyone, can think in a deluge of data is an ongoing formidable challenge.

Perhaps more than anything else, genomics—the digitization of DNA—has illustrated the limits of assuming the democratic and ethical value of making data public.[23] As we saw throughout this book, the goal of making genomic information public fueled investments in the field. During the sequencing of the human genome, scientists, politicians, and journalists alike promised that the genome was a thing of great value to all people: it was "the secret of life," "our common heritage," and the key to a revolution in medical care. Thus it should be in the public domain and not controlled by a powerful corporation. This powerful moral story fueled the race to complete the human genome.

Now, more than a decade later, despite the failure of the greater visibility and circulation of genomic data to create much of medical value, the drive to free data from institutional silos not only drives investments in genomics but also biomedical informatics more generally. In the winter of 2016, US Vice President Joe Biden, tapped by President Obama to lead the National Cancer Moonshot Initiative, called on researchers to "break down silos and bring all cancer researchers together."[24] Today, all other values—privacy, informed consent, inclusion—give way to this effort to free information, this time from our own flesh and blood. This is so despite cases like Generation Scotland that illustrate that the value of these new public biomedical informatic infrastructures is far from clear. The data and tissues may be available, but the route to collective goods remains largely uncharted.

Nonetheless, support for an informatic approach to biomedicine—of which genomics is emblematic—remains strong. The ability of these efforts to frame sharing data as a moral good continues to propel them forward. This is true even though from the start bioinformatic notions of publicness fueled private industry—from the ABI monopoly to the sequencing companies that today benefit from using the Personal Genome Project's public genomes. At its heart, bioinformatic infrastructures are built around the values of business: speed, efficiency, growth. Public genomic data today serves these ends.[25]

Lifeforce, Salesforce

Perhaps nowhere is this more evident than in this city by the bay in which I have lived this last decade. In the fall of 2015, a luxury cruise boat, the *Celebrity Infinity*, moored in the San Francisco public harbor to provide accommodation for the 160,000 people who arrived to take part in the largest technology conference in the world: Dreamforce, sponsored by Salesforce. Mark Benioff, CEO of the Fortune 500 company Salesforce, and namesake of the new children's hospital in Mission Bay, began his keynote at the Moscone Center by acknowledging his mother, a breast cancer survivor, and by highlighting the work of Laura Esserman, a UCSF professor who leads the Athena Breast Health Network. Athena, named for the Greek goddess of wisdom, aims to enroll 150,000 women in California to collect information—including genomic information—to determine if a "personalized" approach to breast health care would be better than standard public health recommendations. Esserman explained the value of working with Benioff and his Salesforce team: what works to know and communicate with customers, she argued, works to know and communicate with patients. Benioff, in turn, explained the value of working with Esserman: precision medicine paves the way for precision enterprises.

In this new world, business and biology unite forces. Lives will be saved through proper management of data in clouds. The political economic power of those who create the data systems recedes from view, as the care of a son for his mother takes center stage and becomes the rallying call for collective action. We are all in it together, Benioff and Esserman explain. "Everyone," Esserman told the crowd, "must play a new role. . . . All the people, the patients, step up to the plate! Share your data." After all, why not? Who would not want to save their mother from dying from cancer while taking part in a revolution? "I am confident that with your [Benioff's] backing we can change the world," Esserman concluded.[26]

With the goal clear—a better world for patients and consumers alike—Benioff and his co-founder Parker Harris focused the remaining hour and a half of their keynote on how to achieve it through their new Salesforce platform. This platform mines Internet events—sending and receiving e-mail, scheduling meetings, and such—for data that can help companies better manage their relationships with customers. While presented in rock-concert-style fashion—Harris dressed up as an action hero, lightning bolt in hand to represent the speed of the new platform—the content delved into procedural details. Harris and Benioff explained how the Salesforce platform visualizes data analytics and enables companies to better communicate and connect in the marketplace. A better world through better management.

This quick shift from inspiring goals and purposes—saving lives—to management—creating platforms, algorithms, and data analytics—also characterized genomics over this last decade. While every initiative described in this book started with a grand vision of purpose—from revealing the secrets of life to countering racism to democratizing science—ultimately, the vision receded and the focus moved to creating *anything* of value from genomic data. In Scotland, an effort to collect the genomes of Scots began with vigorous discussions about social contracts and benefit sharing but today focuses on keeping the samples alive through generating enough funds to keep them on dry ice. In Camden, New Jersey, the Coriell Institute may send out HapMap samples, but community advisory groups no longer meet.[27] On West Evelyn Avenue in Mountain View, California, 23andMe still develops algorithms, but the company faces a class action lawsuit that claims that 23andMe test results are "meaningless."[28] After years of being told about the great value of genomic data, the result for many citizens who have taken part in genomics is that either their DNA was not used (as in the case of Tuskegee) or it is stored in a biobank where proper administration has become the overriding concern.

Yet of course genomic data is not just meaningless and merely constitutive of new forms of administration. One quick bike ride down Third Street in San Francisco confirms this. The ability to produce ever-increasing amounts of genomic data at an ever-cheaper cost point today animates visions of so-called precision medicine. This, in turn, brought to the Mission Bay, Bayer, Merck, and GlaxoSmithKline, who speculate that new lucrative assets might soon be theirs.[29] With them came hundreds of new biotech start-ups. The buildings they occupy exude grandeur worthy of the brave new worlds they promise.[30] Venter may tell us that the human genome sequence revealed nothing, but his fellow scientists and citizens are not fooled: there is something here that is alive, vital, and powerful. Yet what capacities do citizens and societies have to evaluate the gleam and to determine whose lives are brought to life in these feats of architectural and biomedical wonder, and whose are forgotten in their shadows?

Undoing the Biopolis: Public Relations as Public Dialogue

The opening decade of the new millennium was marked by numerous experiments across the globe to build such capacities. The NIH spent millions on developing robust public engagements for the International Haplotype Map Project. The Scottish government invested heavily in developing democratic deliberation for Generation Scotland. However, in recent years, support for public dialogue has transformed into support for public relations. Efforts to launch "precision medicine" in the United States exemplifies this trend. When in 2013 then chancellor of UCSF Susan Desmond-Hellman sought to position her university as ground zero for this new so-called revolution in medicine, USCF hired a public relations company to work with them to launch a MeForYou social media campaign. The goal was not dialogue, or even solicitation of DNA samples and medical information, but rather meaning management. Visit the MeForYou website then and today and it is unclear what MeForYou is for other than to interpolate its visitors into a vision of genomics and precision medicine as an unquestioned good that will save children's lives. As UCSF spokesperson David Arrington explained: "We're trying to put a new thought into the public consciousness of, medically speaking, what can they do for someone else?"[31] The answer, as Esserman explained at Dreamforce: Share your data.

Communication with patients changed over the decade to foster a new PR approach to managing the subjects of research. While at the beginning of the decade Generation Scotland went through doctors' offices and institutional review boards to contact potential research subjects, later in the decade 23andMe created algorithms to manage recruitment into its research arm, 23andWe. At the end of the decade UCSF partnered with public relations professionals and Salesforce to promote precision medicine. For many, these developments did not represent a corrosive intrusion of corporate culture into the protected space of the doctor-patient relationship, but rather innovation. Rather than a critique, UCSF received an award. The Council for Advancement and Support of Education deemed its MeForYou social media campaign the most "innovative use of technology."[32] Universities and companies alike increasingly use the techniques and media of the sales pitch to communicate with patients. Billboards imploring citizens to support genomics and biomedicine spot the highways.

This form of communication differs markedly from HapMap community engagements and Generation Scotland focus groups. Instead of understanding what people want and believe and then attempting to create practices and policies that align with the will of the people, today's precision medicine initiatives begin with the policy—share data—and then attempt to "put" ideas into citizens' heads that will make them want to comply. There is little space for critical discussion. There is little if any space to ask who benefits. This time the answer is given: we all will if we give our DNA and data. This, we are asked to believe, is how a new social contract with biomedicine forms.

Increasingly, this approach rings hollow and creates dis-ease. Some clearly benefit: the researchers and business people who enjoy the beautiful architecture, landscape, gyms, and massage facilities of the new UCSF Mission Bay campus on upper Third Street.

Author playing on interactive digital screens in the lobby of Benioff (photo by Lindsey Dillon).

The benefit for others, for example, those living on lower Third, is far from clear. How can we discuss and respond to these massive transformations in biomedical and informatic infrastructures and how they differentially affect citizens? I offer here a few modest proposals.

Acting beyond the Limits of Liberalism

An essential first step will be recognizing that the practices, meanings, and effects of the central principles and values of Euro-American liberal democracies have undergone radical change. Information is at the heart of the transformation. Postgenomic initiatives took shape over the same period that Google, Facebook, and Amazon became major players in global capitalism, transforming information into a raw resource and medium of capital formation. It is no longer possible—if it ever was—to imagine that the institutions and practices that create information sit outside of power, helping citizens to make democratic decisions. Instead, they are conduits and centers of economic and political control. Thus, it should not surprise that postgenomic calls for more open flows of information and more robust informed consent procedures failed to provide the grounds of public trust or ethical action. As the cultural critic Jodi Dean astutely observed at the beginning of this millennium: "No one today should accept a model of political life that would work just as well as a motto for Microsoft or AT&T."[33] Or Facebook or Google, to update Dean's formulation.

What should provide the ground for enlightened democratic societies in a moment when information proliferates beyond most citizens'—and, indeed, scientists'—capacities to understand its meaning? What principles and practices constitute just governance when the central pillars of liberal democracies—access to information and participation—generate the labor and raw materials for twenty-first-century bioinformatic capitalism?

It is all too easy to continue to fall back on old models and to call for reform. Witness what is currently happening in the United States. At the moment, the US Department of Health and Human Services is considering significant changes to the federal policy for the protection of human subjects known as the Common Rule. An article published in the *Lancet* in September 2015 describes this effort to "modernis[e]" the Common Rule as crucial to the Precision Medicine Initiative.[34] Central to the proposed changes to the Common Rule is the requirement to obtain informed consent for secondary uses of biospecimens, regardless of whether the samples are identifiable.[35] This would effectively prevent the situation I encountered at my UCSF doctor's office. Now I would have to give my informed consent. I would not be denied medical care if I did not want my biospecimens and data incorporated into UCSF research infrastructures. Problem solved?

Unfortunately, no. This change may alleviate the immediate feelings of injustice created by forms that force citizens to give up control of their biological tissues and data if they want medical care. However, they also may inadvertently do more harm than good by seemingly giving people the power to grant consent for use of their data and tissues while failing to create the conditions in which this consent is meaningful. This is so first because the changes would ratify what researchers in Tuskegee worried about: broad consent. If adopted, DNA and data collected for one purpose could be used for all future uses without requiring an additional consent process. This would effectively remove the possibility of citizens controlling the use of their DNA and data for some purposes and not others. Yet social science and bioethics research, as well as the engagements with citizens described in this book, clearly indicate that many citizens want to retain some control over the use of their bodily tissues. Indigenous and Native peoples have time again insisted on it and have created legal frameworks to implement it.[36] As we will see later in this conclusion, they are not the only ones asking for this kind of control.

Second, the proposed changes instruct researchers to shorten informed consent forms so that they only include "appropriate details" and "information a reasonable person would want to know."[37] However, they do not specify how this might be done. How to shorten forms in the face of an increased number and complexity of issues raised by data and tissue use is far from clear.

My analysis of the many efforts made over the course of the decade—from Tuskegee, Alabama, to Ibadan, Nigeria, to Glasgow, Scotland—to gain the input of citizens reveals that many want information about who will benefit from genomics and informatic approaches to biomedicine and health care. In the United States, Scotland, and Nigeria those asked to take part in genomics research asked who will benefit from the research in the broadest sense, not just who will receive medical treatments, but which persons and institutions will receive the jobs, profits, research, and recognition created by considerable public and private investments in genomics and bioinformatics.[38]

It is critical that we make visible how money flows, buildings are built, resources are allocated, research priorities are set, and energy is consumed as Fortune 500 companies and universities invest in information about our bodies and genomes as the route to health and justice in the twenty-first century. What, for example, is the energy rating of UCSF's Athena trial?[39] How many of the Earth's dwindling resources do we use as we ask what the value of breast cancer SNP data is for breast health? In the year 2000, to sequence just one human genome Craig Venter needed Celera's local power company Pepco to put in a new transformer and power pole.[40] In 2010, at my own university, processing genomic data reportedly used up all of the campus's spare energy capacity.[41] In 2015, biologists warned that the computing power needed to manage genomic data soon would exceed that of YouTube and Twitter.[42] The monetary costs of genome sequencing may come down, but as big data projects rival air travel as the number one consumer of carbon on the planet its environmental costs will only go up.[43]

Despite dreams of endless progress and openness, we live in a world of constraints. No matter how powerful or "smart" our computers, not all questions can be answered. The PGP chose to make the final calls, and not the raw data of its genome sequence reads, available. The NIH invests in "improving the health" of "African populations" through funding genomic research, not through funding research on community approaches to health care.[44] UCSF invests billions in precision medicine but plans to close local community health clinics. Major biomedical funding bodies direct their resources toward tailored medicine. As in the world of fashion, tailoring is an expensive affair that does not include us all. While genomics may provide leads to new drugs, those drugs will not come cheap.[45] Difficult decisions confront health care providers and governments as they attempt to decide how to properly allocate public and private resources to create access to the so-called molecular medicine.[46]

These decisions matter greatly. They determine the ends and purposes of today's growing bioinformatic and biomedical infrastructures. They shape the constitution and meaning of things that powerfully gather and direct our lives in these opening decades of the twenty-first century: personal information, health, global health, biomedicine. As such, they deserve our collective attention. However, only rarely over the course of the decade did they become the subject of discussion in postgenomic efforts to inform people and include them in decisions. Those initiatives that did make the greatest effort—the International HapMap Project, Generation Scotland—did so at the beginning of the decade. By its end, these efforts either had been forgotten or recast as naive, and even counterproductive. Bioethicists, policy makers, and social scientists no longer talked about community engagements as the leading edge of ethical engagements with research subjects. Leaders of GS viewed their effort to create a resource of, by, and for the people of Scotland as shortsighted.

Yet good science, as the science studies scholar Charis Thompson argues, requires that we ask these questions. Good science and the good society are made together. Any modicum of a socially, politically, and environmentally robust postgenomics will require practices for knowing and governing that move far beyond the narrowly circumscribed powers of informed consent to develop the collective skills and practices that can respond to the more fundamental questions. What are the things that we should gather around and constitute as public goods? What is the place of genomics, biotechnology, and biomedicine in constituting these goods? How do we decide? Who are "we"?[47]

Justice and the Arts of Collective Judgment

Answering these questions will require long-term investments in institutions that support the *arts of collective judgment*.[48] I refer here not to legal judgments about whether a law has been broken or to scientists' judgments about the results of an experiment.[49] Rather, I am concerned with prior judgments about what laws should govern us and what research questions should inter-est us.[50] These judgments lie at the heart of justice. They call us to attend to the first principles that guide any society and its pursuits of knowledge. John Rawls captured this aspect of justice in the opening pages of his enormously influential book, *A Theory of Justice*: "Justice is the first virtue of social institutions." No matter how efficient or well ordered, Rawls argued, if the laws and practices that form institutions are unjust, they must be redone.[51]

These more fundamental questions about how we should know and live in the world in the midst of the rise of informatic capitalism, growing inequalities, and intensified processes of biomedicalization lay barely beneath the surface of all the genomic initiatives discussed in this book, ultimately troubling and fundamentally shaping their form and ends.[52] While no one doubted that sequencing became significantly cheaper and more efficient over the course of the decade, questions repeatedly arose about how to judge the moral valence of these advances. During the Human Genome Project era, the human genome sequence became an emblem of a struggle for justice in which public scientists fought off private efforts to enclose it. Many celebrated the Bermuda Accords' principles of sharing as central to these efforts to secure the human genome sequence as a public good. Yet at the same time, some worried that infrastructural changes to gene mapping and sequencing ushered in by genomics threatened to undo long-cherished communal practices of sharing. Most poignantly, in his autobiographical account John Sulston described the ways in which genomics put data in the open while it placed human beings in a prison. The very architecture of the new buildings built for genomics, he argued, located sharing of data between machines at the heart of the endeavor, displacing the sharing of ideas among flesh-and-blood humans.[53]

These worries about whether genomic investments in the infrastructures of data and machines might inadvertently displace and devalue thought and life remained an ongoing and defining concern of the post-genomic era. Even practices forged to recognize diverse modes of thinking and living (for example, HapMap community engagements) proved limited in their capacity to recognize concerns that moved beyond the administrative need for informed consent. When broader interests did arise—the constitution of citizenship and health care and the fair distribution of public resources—all involved worked hard to respond. However, ultimately they faced limits. They received funding from research institutions that supported projects with timescales and remits too limited to respond to deeper constitutional issues.

Nor could they address questions about the proper constitution of science and medicine. Should a mode of doing research so dependent on speed, technological innovation, and venture capital dominate the life sciences? Should a field that promised future—not immediate—improvements in health care move to the heart of biomedicine? Natural and social scientists alike posed these questions. "Genomics," Harold Varmus, head of the National Cancer Institute, argued in 2010, "is a way to do science, not medicine."[54] Given this, on what grounds could NIH, Wellcome, and other funders who sought primarily to improve health and medicine justify major investments in genomics?

All of the initiatives described in this book faced challenges as their ability to open up these fundamental issues to collective judgments moved beyond the narrow mandates of the research grants that supported them. This is despite the great efforts made over the decade by genetic counselors, social scientists, and policy makers who committed themselves to responding to community concerns. Those who designed HapMap community engagements did initially seek to let HapMap communities judge whether their communities should participate, but ultimately NIH leadership decided if sampling would move forward. The dialogue groups in Tuskegee, Alabama, came up with several policy recommendations—including the clear statement that "basic health care should come before genetic services"—but the project did not have the support it needed to interface with policy makers and act on these recommendations.[55] Social scientists, lawyers, and scientists working on Generation Scotland devoted considerable energy to community consultations, but once the collection of samples started and issues arose, the money had run out for public-engaged responses. Instead of carrying forward the important work of each of these endeavors to create collective judgments, the decade ended by defaulting back to a particular vision of openness. Genomics should be open to us all, the Personal Genome Project argued. It is a fundamental good to which we should all have access.

How Do We Speak in the Open?

While inspiring, such a stance can too easily foreclose debate and collective decisions. Personal Genome Project board member Michelle Meyer learned this when she sought to argue that agreeing to put her genomic data in the open did not mean anyone could now use her DNA and data in any way they liked. In the eyes of too many, Meyer worried, the opportunity to discuss proper use of her DNA and data closed once she agreed to put it in "the open."

Instead we require greater public understanding and recognition of a fundamental property of openness: no matter how free or open anything might be, decisions must be made. Decision derives from the Latin root *cis*, meaning to cut or kill. Not all forms of life or uses of data can be supported. Some live and prosper. Others fail to gain support and die. A judgment must be made. The question is not whether to judge, but how. As Arendt reminds us, there is perhaps no question more fundamental to political life than this one. To illustrate, in *The Human Condition* she returns to Aristotle and the ancient Greeks for whom "to be political, to live in the polis, meant that everything was decided through words and persuasion and not through force and violence."[56] Do we make decisions through the sword or the word? Much depends on how we answer this question.

For Arendt, the answer was clear: either we learn to speak to each other, or we die. Speech, she argued, is what makes "human togetherness" possible. It is through our efforts to find the words to tell our stories that we reveal ourselves to each other. This speech differs from "mere talk." The latter, talk, is instrumental, designed to fool a foe or dazzle with propaganda. The former, speech, is revelatory, designed to foster relationships and collective life.[57] It requires the presence of others. It is through speaking in the presence of others that a common world forms. Speech, in this sense, is a form of articulation. It joins.

For Arendt, such speech also creates the grounds for judgment. It fosters thinking with others, what Arendt described, referencing Kant, as an "enlarged way of thinking" that is fundamental to judgment. And "judging," she writes, "is one, if not the most, important activity in which this sharing-the-world-with-others comes to pass."[58] Thus Arendt formulates her simple but powerful call at the beginning of *The Human Condition*: "What I propose, therefore, is very simple: it is nothing more than to think what we are doing."[59]

Yet today how we might think that which we do is far from a simple matter. In times in which more and more of life is constituted and conducted via personalized media and infrastructures that create echo chambers of our own thoughts, how might it be possible to speak and think with others across our differences?[60] As Arendt constantly asserted, human plurality is "*the* condition of political life."[61] How do we gather in our plurality to think what we do in the postgenomic condition?

Reclamation of the Public

Early on, leaders in genomics expressed hopes that genomics might help with this critical political task; it would gather us in our plurality, and not divide us. At the celebration of the completion of the Human Genome

Project, Craig Venter announced: "We have sequenced the genome of three females and two males, who have identified themselves as Hispanic, Asian, Caucasian or African American. We did this . . . to help illustrate that the concept of race has no genetic or scientific basis."[62] Despite our differences, Venter and other leaders of the HGP argued, we are all one, and the genome is for us all. This framing played a critical role in enabling the HGP to become a crowning achievement of the public governments of the United States and the United Kingdom, and for the human genome sequence to emerge as a quintessential public good of the opening decades of the twenty-first century.[63]

However, over the course of that same decade, decisions about genomics took place increasingly out of public view in spaces that were less accountable, more homogenous and hierarchical. At the decade's start, planning for the International Haplotype Map Project involved several meetings designed to elicit the input of diverse communities—from indigenous rights organizations to patient advocacy groups. At the decade's end, the rollout of the Precision Medicine Initiative (PMI)—the United States' national effort to analyze the genomes of one million Americans—came direct from the Office of the President. In January 2015, President Obama announced the PMI in his State of the Union Address. According to NIH Director Francis Collins, NIH was then given only nine months to come up with the design. It was, he argued, a "very fast moving process" that put "the pressure upon all of us."[64] The Director's Office of the NIH convened a working group made up only of biomedical and health care experts from academia and industry. The working group organized four workshops in late spring and summer and issued a report in September 2015 that laid out key challenges and recommended ways forward.[65]

Much like Laura Esserman's Athena clinical trial and UCSF's MeForYou campaign, the PMI assumes that genomic approaches to medicine have clear public value. As the president's chief data scientist DJ Patil wrote in a blog post in August 2015, the PMI would lead to a "new era of medicine." In it everyone will receive "tailored treatments" and "become an active participant in scientific discovery—furthering an open and inclusive model" of biomedical research. To make this public promise of genomics a reality, Patil, like Esserman, argued that we all need to do our part: "We need all sectors to work together. We need people to actively engage in research and voluntarily choose to share their data with responsible researchers who are working to understand health and disease."[66]

PMI comes at the end of now over a decade of efforts to enroll diverse people in the conduct of genomic research. As my account of these efforts makes clear, these efforts faltered as answers to basic questions about whether genomics was a thing of value, and for whom, remained unclear. Given the growing empirical evidence that it may have clear uses in some cases, but is far from creating a medical revolution, it is questionable whether the uncritical promotion of PMI will lead to broad support and trust of the initiative. Indeed, many prominent news outlets have already run commentaries that question its value. Some question the public benefit of a precision medicine approach that supports the development of drugs where "prices in the tens of thousands of dollars per year will become the norm, with certain drugs commanding six figures per year." Others question whether sufficient understanding of genomics exists to launch the PMI.[67] Both concerns are not new. As the stories in this book illustrate, they arose across the decade, and across the globe.

Rather than an emblematic public good, genomics today bears witness to the broader erosion of the meaning of the public as a domain that fosters consideration of common concerns and the creation of collective goods. Instead, PMI's invocation of the public is part of a public relations campaign designed to enroll people to give of their personal data and DNA. Its public is not a space designed to engage in the difficult work of creating common meaning and valued things, but rather one filled with words meant to dazzle and enroll.

If genomics is to make good on its claims to provide public goods of broad value that support a diverse range of lives, its leaders must respond to these transformations in the meaning, value, and enactment of the public. Any invocation of genomics as a public good increasingly will ring hollow in the absence of spaces in which a broad range of citizens gather to speak with one another about its fundamental ends and purposes.

Genomics: A Matter of Public Concern?

How to create these spaces is of course a formidable challenge. In the United States, there is more than two centuries of debate about the viability of civic institutions. From Tocqueville to Robert Putnam, observers of American society have argued that the country's individualism and the breakdown of bonds between individuals and collective life threatens the viability of public life.[68] Other scholars diagnose the problem as a broader one: it is not the problem of America, but of modernity.[69] Still others disagree that there is a problem: civic engagement has not declined; it has merely changed form. People may no longer go to church or join bowling leagues, but they do create spontaneous public spaces and occupy established ones using the power of social media. Witness the protests in Tahrir Square and Occupy Wall Street.

Whether in decline or merely transformed, what is not in question is the importance in democratic societies of cultivating public culture and civic institutions that gather citizens in the presence of others to speak about matters of concern. The challenge is whether and how these spaces come to be. When, for example, bioethicists and genetic counselors sought to gather African Americans from across the economic spectrum in Tuskegee to discuss genomics, many refused. What, they asked, did genomics have to do with their lives? When offered $25 to give a blood sample, many consented. The message was clear: money for a next meal matters; genomic medicine does not.

As the decade went on and investments in genomics intensified, many did begin to feel that something of great consequence was underfoot. However, as the story of the efforts to gather people for the HapMap community engagements make clear, people took part not because they thought that genomic medicine would improve the health of their communities anytime soon, but because they believed genomics had great symbolic and economic importance. Genomics mattered to those in power who represented national governments and large corporations. Thus being a part of it, some believed, might be a way for their communities to gain recognition and support. For robust civic spaces to form around genomics, these more fundamental questions about the role genomics can and should play in constituting and recognizing citizens must come to the fore. This will require that we redefine possible ways to think and enact genomics.

Redefining Possible

Redefining Possible. This is the slogan of the new UCSF Benioff Children's Hospital that today spots the San Francisco highway scape.

For UCSF, "redefining possible means state of the art technology and compassionate care." It means harnessing "the special sense of urgency

Billboard along Highway 101 in San Francisco (photo by author).

that comes from caring for children" to the task of creating technological breakthroughs.[70] This effort to link the future of children to technological breakthroughs marks not just the Benioff Hospital but also UCSF's whole precision medicine MeForYou campaign. The MeForYou website features a young girl, Georgia, who reportedly has a slight risk of triple negative breast cancer. She is why UCSF does what they do: "For Georgia."[71] At my own university, children also are the focus of a marquee project of our new Genomics Institute.[72]

Sharing data to try to save children's lives is a public good that supporters of genomics hope many will gather around. However, the stories of the postgenomic condition call us to gather around so much more. We must redefine possible, including what it is possible to imagine when we endeavor to redefine possible. As Arundhati Roy implored in a lecture she gave in the Science and Democracy Lecture Series at Harvard in 2010, we must "stop making war on those who have a different imagination."[73] At the moment, the institutions, capital, and Salesforce that seek to create one data platform for customers and patients define the arts and technologies of possible within genomics and biomedicine. To raise any questions about this administrative goal is to risk being framed as against health, against democracy, and ultimately against children. This is a narrow vision of the future that forecloses too many urgent concerns. This is plainly evident once one leaves the Benioff Hospital and walks only a few blocks farther down Third Street. Extreme poverty, lack of access to basic health care, gun violence, nuclear contamination. Surely these also are important to the future of our children.

The problem of meaning that lies at the heart of the postgenomic condition arises from an overly narrow biotechnocratic understanding of life, health, hope, democracy, and the future that eclipses these broader needs of people and the planet. Time and again, when consulted by genome scientists and policy makers, people from across walks of life and political and economic worlds attempted to articulate genomics to these broader concerns of their daily lives. Those in Tuskegee spoke of lack of access to a hospital. Those in Ibadan, Nigeria, initiated discussions of support for their community health center. Those in Houston, Texas, envisioned an endowed chair at a local university so that their ways of knowing and living in the world would be passed on to future generations of the Gujarati diaspora. Those in Scotland pressed for sharing the benefits of genomic research with the people of the Scottish nation. Through the dedicated individual efforts of the government employees, scientists, bioethicists, anthropologists, and sociologists who worked on these genomic initiatives, in some cases concrete responses resulted. NIH provided support for expanding the hours of the community health center. Generation Scotland created an agreement about benefit sharing. However, in each case the gains were limited and short term. While NHGRI did arrange for support for the community health center in Ibadan, Nigeria, they resolved never to provide that kind of support again. As a bioethicist involved in the deliberations explained to me, it was not feasible to offer this to all communities; NIH is in the business of research, not health care. As other genome scientists explained, while they believe allocation of resources and meeting basic health needs are important, these are not things they have the power to change.

Yet, is this true? Surely, a different response is possible. For this, though, we require different stories.

Re-membering Life

Stories, Arendt argues, are the end result of speech and action; they arise from our life with others. They reveal us and make our thoughts and actions real and meaningful to others.[74] They gather us, as they are of interest; that is, they are that which lie "between people and therefore can relate and bind them together."[75] They also help us to re-member—both in the sense of facilitating an ongoing-ness of memory, and in the sense of reworking who we consider "kin and kind" of our common world.[76]

This book is one modest contribution to the effort to create stories that might gather us to create collective meanings and worlds that respond to plurality. Through the stories I have told, I hope that readers have experienced being in the presence of others whose ideas about genomics may differ from theirs. From engagement with these differences, I hope new grounds of thinking with others, and thus judging collectively in post-genomic times, will arise.

A recent effort in the San Francisco Bay Area demonstrates just this role stories can play in gathering diverse people to debate and judge genomics. In 2012, the National Cancer Institute (NCI) funded the Alameda County Network Program for Reducing Cancer Disparities.[77] Alameda County is home to a high percentage of people who are uninsured or underinsured. The NCI sought to increase participation of these underserved communities in cancer screening and biobanking. The Cancer Genome Atlas (TCGA) had just started, and one problem it faced was the familiar one of low participation of minority populations.

Julie Harris-Wai, an assistant professor in the Department of Behavioral and Social Sciences at the University of California, San Francisco, and associate director of the Center for Transdisciplinary ELSI Research in Translational Genomics, did not want to recruit for TCGA, but did want to learn from underserved communities about their concerns about biobanks. Thus, in 2013 she applied for supplemental funds to support dialogues with African American communities in Alameda County.[78] Specifically, she proposed gathering together to discuss Rebecca Skloot's *The Immortal Life of Henrietta Lacks*. In partnership with African American churches in Alameda, Harris-Wai and community leaders convened reading groups to discuss this story of the extraction of tissues from a young African American woman without her consent and the use of these tissues to create what some herald as the first biotechnology, the HeLa cell line. Harris-Wai reported that the book acted to bring people together. The discussions were wide ranging and explored critical concerns: black bodies are the subject of research, but not of medical treatments; scientists are enriched while communities are impoverished; communities lack access to meaningful information about the practices and goals of research. Discussion of these issues led to exploration of possible modes of redress. A common refrain emerged: "I won't participate in the research until I can see something for my people."[79] This is not, however, where the story ended. Members of the reading groups also proposed specific actions: "Simply give us a number, just like a Social Security number, a driver's license number . . . that we can log onto a database, and we can know everybody who's accessing our specimen, and we can know exactly what our specimen is used for."[80] Such a tracking number, others suggested, could be used to compensate individuals and communities for use of their samples.

In addition to making it easier for communities to understand and respond to research, the reading group participants argued that health professionals needed to better understand and respond to communities. Health care professionals, they argued, needed to slow down and educate themselves about the people with whom they worked. Only then would they be able to effectively connect to and work with communities.[81]

Gathering around stories—as theorized by Arendt, and enacted by the Alameda churches reading groups—cultivates thinking across differences.[82] This enlarged thought provides the grounds for judgments about how to create worlds that support and foster more diverse lives.

Justice: Creating the Institutional Conditions of Thought and Life

Yet how can it be sustained? The church members and researchers who came together in Alameda confronted a similar challenge to the one that bioethicists, genetic counselors, and community members faced a decade earlier in Tuskegee: they received support to conduct the dialogue but not to translate the outcomes into policy change. Further, the overarching goal from the point of view of the funding body was not to engage communities in a discussion about the value of genomic research and biobanks but rather to increase rates of minority participation. The broader discussion happened because of the unique training and orientation of the principal investigator, Julie Harris-Wai, who has a master's degree in public health in health education/health behavior (specializing in community health promotion) and a PhD in public health genetics. For the work to be ongoing, and to lead to substantive change, not just individuals but institutions must take on these commitments to understand the value and meaning of novel forms of technoscience for diverse lives. For this to happen, I argue that justice must become a gathering concern.

Justice, of course, is not new to reflections on science and technology. As I noted earlier in the book, it along with informed consent and beneficence were the three principles put forward in the Belmont Report, the defining document for bioethics, at least in the United States. However, as the legal scholar Patricia King pointed out in her 2004 oral history, she and the other authors of the Belmont Report did not have "too much of a clue" about what to do with justice. At the time, it did not seem urgent. The focus was on "Tuskegee-like" problems in which research was being done on human beings without informing them or asking for their consent. Informed consent appeared the obvious solution.[83]

Today, however, in a time when access to information is widely promoted, in some ways the problem is just the opposite. There is too much information. What is its value and for whom? Rather than push for open access to all information, what is needed is broad engagement with questions about how information itself forms—for what ends, and for whose benefit? These questions about the formation of the first principles—or, *à la* Rawls, the first virtues—that govern our collective actions are questions of justice.[84] How to answer them is not straightforward. While the task of informing a research subject about what a research project sets out to do is far from an automatic task, it begins in obvious places: research proposals; human subjects review applications. It is less clear how to evaluate the formation and value of information itself. It is not something that conforms to ordered bureaucratic procedures. Instead, it requires engaging with people—many people. It requires attending to differences. What information might mean and what benefits it might produce for some may very well be different from what it means for others. Our efforts to categorize those differences based on human groupings often will fail. As those who worked on the Communities of Color and Genetics Policy Project emphasized, it is not possible to speak about the "African American" view of genomic information. There are many differences that arise from different class, race, gender, religious, and historical backgrounds. A similar critical sensibility must accompany our understanding of who "the people" are. They are not, I argue, opposed to scientists. Scientists, too, are not homogenous. They also differ in their assessments of the value of genomic information.

Creating the institutional conditions in which scientists, humanists, artists, community members, activists, and policy makers can discover and engage these differences is critical to creating the understanding needed to produce decisions that many might deem just. Over the last decade at my own university, I have been part of collective efforts to cultivate formal and informal infrastructures for generating these encounters. From the very start, these efforts included our genome scientists as well as natural and social scientists and artists from across the university. A focus on justice proved critical. Early on we formed the Science and Justice Working Group to collectively address problems that might concern us all, but for which we brought different approaches and perspectives.[85] Art practice, anthropological methods, and the material sciences came together to produce a greenhouse that became the hub of understanding what sustainable food production meant to different communities of scientists and farmers. Social scientists and historians worked together with forensic anthropologists to help create a dialogue about the use and meanings of racial categories within the profession and practice of forensic science.[86] Crucial to our success has been ongoing institutional support. While our efforts received support from the National Science Foundation, they would not be possible without the ongoing support of the central administration of our university.[87]

Meaningful and transformative engagement with questions of science and justice require this kind of ongoing institutional commitment. How might it be achieved beyond the bounds of a university in the redwoods of Northern California, known for fostering novel thought?

While I cannot venture to answer this question writ large—answers will be many and will arise out of specific contexts—in the United States I argue that the time is right to revisit the Belmont Report and to convene a new national commission to focus on the principle of justice. This commission could at minimum address the following three questions. First, how can publics form and gather to foster collective judgments about the value and meaning of science and technology? To answer this question, the commission could review the efforts made to date to engage and consult publics on emerging forms of science and technology. From Tuskegee, Alabama, to Edinburgh, Scotland, to Ibadan, Nigeria, from genomics to stem cells to nanotechnology, what have we learned? While the NIH, Wellcome, and other national funding bodies have invested in consultations and engagements, as far as I am aware, few resources have been invested in studying and evaluating their outcomes.

Efforts in other domains to understand the formation of public space and critical dialogue also will be germane. Scholarship in urban sociology and geography on the racialized, classed, and gendered constitution of these processes will be particularly pertinent. As those involved in the Communities of Color and Genetics Policy Project explained, it matters where "democratic deliberation" takes place and who facilitates it. "Some individuals may feel comfortable coming to a hospital or government building and having the conversation led by a public health official, for example. Others may feel more comfortable meeting in a church, while others may rather meet in a secular community center."[88]

The second question that should be addressed is, "How can public dialogues link to policy change?" As Arendt constantly reminded, speech cannot be divorced from action. As many indigenous rights activists and scholars asked to participate in dialogues about genomic research repeatedly argued, speech without action is another form of oppression; it requires labor but delivers no enduring change.[89] Initiatives such as the Precision Medicine Initiative should build in public engagements that clearly link to mechanisms for changing policies. When the Human Genome Project first launched, there existed in the United States an Office of Technology Assessment that could help facilitate these links. It closed in 1995 as part of a conservative critique of public government.[90] Twenty years later, it is time to revisit the need for an institution that can support critical public deliberation of science and technology.[91]

Finally, the third question: What is the relationship between innovation in science and technology and inequality? The growing gap between those who have and those who have not is perhaps the defining issue of the first decades of the twenty-first century. It also arose time and again as people gathered over the decade to make sense of the meaning of genomics in their lives. What sense does genomic research make when basic health care is beyond their reach? This question deserves our collective attention.

Concerted focus on these issues, as well as a call for funding bodies and ethical review committees to attend to these issues, could go a long way toward beginning to have a clue about what to do about justice.

Biological Thought in Posthuman Genomic Times?

A focus on questions of justice, I argue, also will help facilitate attending to the fundamental questions of knowledge raised by genomics. Discerning the value of genomics is not just a question of who benefits, but a question of what we can know. Who can make sense of genomic information to become today's biological and medical expert? Can any person at all understand the human genome, or has genomics ushered in an era where computers, not humans, think?

Today digital computers lie at the heart of genomics. Because of the massive amounts of data involved, many argue that these machines will remain an obligatory point of passage.[92] As one architect of genomic information infrastructures explained to me: You "get to a point with so much data that you need computers to reduce to something understandable."[93] Yet how far should this argument extend? While computers might be essential for big-data whole-genome analysis, this is not the only approach to interpreting genomes. Within the genomics community, there are teams of scientists who work on annotating genomes not through writing computer code, but through analyzing the sequence data themselves. Like the biocurators at the personal genomics companies, they make judgments about where the meaningful parts of human genomes lie—where, for example, genes stop and start—through interpreting genomics data using analytic lenses created from careful reading of the scientific literature. Yet while many in the life sciences recognize the value of their work, increasingly it is hard to support and publish it. As a member of one such team explained to me: "I think the editors actually value automatic prediction better than they do manual annotation. It's hard to get things published, because you only do a subset. They always say, 'Well this is not genome.' We did alternative splicing comparison of 1% of the genome for the ENCODE region, compared it to mouse and human and we found some novel findings. But of course it's not genome-wide so *Genome Biology* won't even take a look at it and send it out to reviewers."[94] In the effort to sequence the human genome, speed and scale became defining values. Thus the field of genomics focused on removing slow-moving humans from the scene, replacing them with faster, computer-enabled automated sequencing machines.

Now that the task is no longer sequencing the human genome but interpreting it, the bias toward automated approaches—such as whole genome analyses—affects what we can know about genomes and the quality with which we know it. Consider, for example, that more than a decade after the completion of the Human Genome Project the genomics community still lacked an agreed-on consensus about where protein-coding regions are in the genome. Computer-enabled automated annotations that predict their locations are available. However, as one annotator who works on developing the human genome reference sequence explained to me: "We still haven't manually annotated the whole genome first pass. So there still are automated predictions, especially on [chromosomes] 15, 16, 17, and 18." These automated predictions get things right most of the time. However, she explained: "There are always exceptions to the rule. That's why you need manual annotation."[95]

The entrance of personal genomics companies into the field of genomics highlighted these deeper epistemic problems. When Linda Avey, Anne Wojcicki, and Sergey Brin first imagined a company that would enable one to Google their genomes, they imagined bringing Google's computer power and data management and processing prowess to biology. The main problem they highlighted was what they viewed as an outmoded, paternalistic ethical regime that prevented people from accessing the powerful information contained in their genomes. They quickly discovered this would not be the only—or even the main—challenge. More importantly they confronted a problem of knowledge: could algorithms know life? Computers and informatics might provide excellent tools for storing, managing, and looking for patterns in massive amounts of genomic data. However, transforming genomic data into meaningful knowledge still requires humans to make judgments about which algorithms to use and about which data to input into them.[96]

What happens when algorithms replace the training of minds? What then becomes of the judgment needed to know the world? These questions—formulated by Lyotard and Arendt—remain with us. Many of the computer scientists and mathematicians I talked to over the decade hope for a pragmatic answer. As a staff scientist who works closely with the public efforts to annotate genomes explained to me: "Computers should be leading and telling people what to look at."[97] In this imaginary, computers do the rote work that humans should not have to spend their time on and guide the way to the more difficult interpretive problems. In other words, computers free human beings up to work on the more interesting and difficult biological problems.

However, within the broader field of the life sciences, the rise of computers and informatics poses a more fundamental question about the constitution of biology: might computers and informatic techniques not just help biologists do what they are already doing, but change the very nature of biological practice? As the historian of bioinformatics Hallam Stevens has shown, informatics brought modes of work and thought into biology that required scaling up and speeding up data production and management. This mode of work might suit computers and the computer scientists who program them.[98] However, what will become of the vast stores of biological knowledge and practices for knowing life crafted over centuries—taxonomy, descriptive developmental biology, for example—that do not fit easily into this big data approach?

The postgenomic condition is one in which we are experimenting with answers. One lesson to glean from the first years of this experiment—one that cuts across public and private research initiatives, from 23andMe to Wellcome Trust's Human and Vertebrate Analysis and Annotation group—is that institutions must still find ways to support flesh-based reading, thinking, and judgment. It may be expensive and complicated to support thinking in vivo, but life requires no less.[99]

Forging a New Social Contract with Genomics and Biomedicine: From Billboards to Science and Justice

We live in times of great promise and great despair, where no matter where you are standing—at a laboratory bench with a pipette or in a street with a protest sign—there is a palpable feeling that much is at stake. In a remarkable move, leaders in genomics marked this sense of fundamental change not just with grand claims about what we can know about life, but with calls for fundamental societal change. Most notably, co-chair of the US National Academy of Science report on precision medicine, Susan Desmond-Hellman, called for a new social contract with biomedicine. In so doing, Desmond-Hellman drew on a long history of linking flesh and blood to democratic citizenship. The idea that tissue donation plays a central role in the constitution of society dates back to World War II when citizens supported the war effort through donating blood.[100] In his much-celebrated book, *The Gift Relationship: From Human Blood to Social Policy*, British social scientist Richard Titmuss celebrated blood donation as an altruistic act that supported the formation of the post–World War II democratic welfare state. Institutions created in the United Kingdom following World War II, most famously the National Health Service, embodied this moral principle. At the same time, the American Red Cross in the United States promoted blood donation as a form of civic duty.[101]

Altruistic acts of donating tissue endured Margaret Thatcher's and subsequent UK prime ministers' dismantling of the welfare state. The long lines to give blood in New York City after the 9/11 attacks indicated that also within the United States feelings of citizenship remained imbricated with blood donation.[102] When Desmond-Hellman and UCSF ask their patients to share their DNA and data with their medical provider, they tap these feelings.

Yet conditions have significantly changed since World War II. The right constitution of a moral economy both for the exchange of data and for blood is no longer clear. Just as scientists no longer share data directly with other scientists to advance the goals of a relatively well-defined scientific community, citizens no longer directly donate blood to another citizen in need.[103] Instead, through a series of complex biotechnical and economic practices blood will be disaggregated into various bits (for example, red blood cells, plasma, DNA, data) and then distributed to a whole suite of institutions that defy easy categorization as private or public. What it means to give or take tissues under these conditions is far from clear. Certainly, it does not mean supporting a social democratic state. Does it instead, as social scientists Catherine Waldby and Rob Mitchell ask, mean rendering "the body an open source of free biological material for commercial use"?[104] The convergence of an ethos of open source with market logics has created palpable dis-ease among medical providers, biomedical scientists, and people asked to give of their blood and DNA. Under these conditions, how can anybody know what the public interest is and if it is being represented when one shares their data and DNA?[105]

Some genome scientists raise similar questions about the sharing of data. Requirements to share can lead to unanticipated problems. The Bermuda requirement to share within twenty-four hours impeded quality control of the data. It also directed many resources toward uploading the data. Ironically, public institutions like the NIH have begun to give up on trying to keep up with providing support needed to maintain all the data. What began as a commitment to public release of data may result in privatized bioinformatic pipelines.[106] Under these conditions, is a commitment to open data still of utmost practical and ethical importance? On what grounds?

For any new social contract with biomedicine to form, these questions about the proper constitution of the moral economies of the life sciences require collective attention. Moral economies built out of the tragedies and hopes of previous eras do not suffice. We must instead redefine what it is possible to think and how it is possible to live together.

Understanding and addressing the conditions of knowledge and justice in the postgenomic era—one defined by massive investments in data infrastructures, the increased reach of biomedicine into the lives of people, and the intensification of globalization and informatic capitalism—requires fundamental new thought, imagination, and resources.

These are not issues that can be addressed in a sound bite and a drive-by, the medium of the billboard. All who live and work on Third Street, from the gleaming buildings to the corroded and polluted docks, require so much more. This is a matter of science and justice.

EPILOGUE

Redefine possible. These times demand it. It is possible.

This I learn partially from the place where I have finished the journey of researching and writing this book: the edges of Tempelhofer Feld. A very large meadow in the middle of Berlin, for centuries it served as a site for military exercises and demonstrations of state power. The Prussian army used the fields for training exercises and to march before Kaisers. The new Nazi government staged a massive million-person-strong rally here on May Day of 1933. Hitler oversaw plans to enlarge the airport (established in 1923) by three times, building what he dubbed Germania, the largest airport in the world that he imagined would serve as the gateway to Europe.[1] His regime moved Konzentrationslager Columbia, one of the first concentration camps it established, north of the city to Sachsenhausen to make room for the expansion. The Weserflug forced labor camp remained.[2] In the end the buildings occupied 300,000 square meters under 1.23 kilometers of curved column-less roof, making it to this day reportedly the largest protected building in the world.[3]

After Hitler's defeat and the end of World War II, the US military moved in and Tempelhof became the main terminal for the American military and the planes that brought food and supplies to West Berlin during the Soviet attempt to cut the city off during the Cold War. The American military remained until 1994. The airport continued to operate until 2008.

Tempelhof then became a site of a struggle. Like Third Street, it was prime property. Developers wanted

Tempelhof Main Terminal Building, August 2016 (photo by author).

A Friday summer evening at Tempelhofer Feld, summer 2016 (photo by author).

to build 4,700 new homes and a public library on 20 percent of the field. They promised some would be social housing. Many Berliners did not believe it and voted overwhelmingly to keep the space for community development by the people of Berlin.[4] Today, against all odds—at least to an American living in the midst of the intense development of San Francisco—Tempelhofer Feld is a space for community gardens, bikers, roller-bladers, picnickers, book exchanges, birds, and community workspaces.

The vote also puzzled some in Germany. "The wonder is that Berlin still carries on," commented Ulf Poschardt, deputy editor of the national

Bike work station on Tempelhofer Feld, summer 2016 (photo by author).

Tempelhofer Feld under a late August sun (photo by author).

daily newspaper *Die Welt*.[5] Yet the people here decided life could go on in other ways. Today Tempelhof is a space where many gather and meet, including the thousands housed in the former airport buildings and hangars.

In 2015, 79,000 people fleeing persecution in their home countries arrived in Berlin, raising urgent housing needs. Some argued that Tempelhof did not have adequate facilities to serve these needs: not enough bathrooms, not enough privacy. Others worried that its development for housing violated the 2014 referendum in which Berliners voted to keep the airfield free of property development.[6] In the end the decision was made to turn its hangars into temporary housing. Today 1,700 people live in these facilities. The accommodations are far from ideal. Mass housing inevitably leads to tensions, short as it is on privacy and the comforts of home.

Yet efforts to navigate these tensions and difficulties have created inspiring responses to the growing xenophobia in Europe. Some have started social network sites that enable established residents of the city to open their homes to newcomers for dinners of shared cooking and learning.[7] Others have developed schemes that easily allow anyone, including an American like myself, the opportunity to volunteer at Tempelhof teaching English, cooking food, and doing mundane tasks like folding children's clothing. Still others have started cafés and bike clinics in the back alleys, where the railroad used to run.

Bike repair workshop, Tempelhof, August 2016 (photo by author).

Tempelhof today is much beloved by many. It is not, I suggest, only because it is amazing to bike and run down an airstrip (which it is!), but because of what goes on in these back corridors, in these efforts to bridge great differences and adversities to make homes for thousands. What makes Tempelhof moving and inspiring is that it is a place where people come together to make another world possible in the presence of great trouble and enormous difficulties.

Third Street, genomics, and biomedical innovation may seem worlds away—both conceptually and geographically. However, there might be cause to read the connections. Germany's policies are unique with respect to immigration and genomics. Direct-to-consumer genetic testing is in effect banned.[8] Immigrants have been welcomed. The reasons are allied. The country experienced the terror of a purportedly rational so-called science-based sorting of its people into those granted the rights of life and citizenship and those stripped of those rights and, in many cases, life itself. Arendt famously claimed the reason was not evil individuals, but a thoughtless bureaucratic machine.[9] Others documented the role that eugenic theories imported from the United States and the United Kingdom played in underwriting sterilization and extermination laws.[10] Some critiqued Arendt for blaming the victim. Many dismissed the problem of eugenics as a matter of bad science. No matter your position, what is clear is the powerful role that claims about genetics have played, and continue to play, in state decisions about citizenship and truth and justice.[11]

It is possible for many to forget about or dismiss these likely uses of genomics in all of the venture-capital-fueled hope for tailored drugs and precision medicine. However, as the stories in this book demonstrate, they were not forgotten by those asked to give of their blood and DNA for genomics in the decade after the sequencing of the human genome. Time and time again, from Tuskegee, Alabama, to Houston, Texas, to Mexico City, Mexico, to Edinburgh, Scotland, the act of giving blood and DNA raised questions of belonging. Am I a valued member of this community, this nation, this world? People across the globe understood that by sequencing the human genome, governments, corporations, and scientists made genomes into important things that mediated answers to these consequential questions. The clear message conveyed as bioethicists, social scientists, and lawyers sought to ascertain and interpret their views was this: these consequential questions about who belongs cannot and should not be forgotten.

These questions are not forgotten in Berlin. The trouble of how we sort self from other is out in the open. It is possible to think and struggle with the problems this brings. It is this, I suggest, that animates the lightness and joy of Tempelhof today. As I finish this book and return to San Francisco, I endeavor to imagine what such a space might look like in that beautiful city by the bay. There, of all places on the planet, other worlds should be possible. I hope that the stories of *The Postgenomic Condition* can aid in their creation.

—Berlin, August 27, 2016

ACKNOWLEDGMENTS

Any power of the preceding words—their capacity to deepen understanding and to foster good in the world—has been enabled by decades of support and countless hours of thought and reflection offered by many people and institutions. There is no way, in just a few pages, to adequately recognize the many debts of gratitude that I have accumulated in order to write a book such as this one. As I thought about how to even try to do so, I found myself returning to the places that brought life to this endeavor. "The only way to find a larger vision is to be somewhere in particular," wrote friend, companion, and creator of so much thought that is good to live with, Donna Haraway.[1] And so it has been true for me as I have created the thought and writing that now lies between these two covers. As this journey ends, I want to offer my thanks to the people and institutions in these places that supported the creation of the vision and thought that formed *The Postgenomic Condition*.

I begin in Kansas, as it was really here that this story began. Early on, beloved friends, family members, and mentors, as well as some remarkably generous senior scientists, believed in a young girl from Kansas who dreamed not about outer space, but rather the inner spaces of molecules, thought, and life. Biologist and marine ecologist Robert C. Worrest, then of Oregon State University, commented in the margins of yellow legal pad pages on the hypotheses and research proposals of a young teenager from the Great Plains who thought she might study the effects of a deteriorating ozone hole on the primary productivity of the world's oceans. Dr. Paul K. Bienfang, of the Oceanic Institute in Waimanalo, Hawai'i, sent me phytoplankton settling tanks. A local chemist, Paul Ruehle, processed my carbon 14 samples. They all launched my life in the biological sciences and demonstrated to me at a young age the power of the moral economies of science at their best. They and my mentors at the University of Kansas—in particular Sally Frost-Mason and Ken Mason, who steered me into molecular biology, and ultimately genomics—taught me a deep respect for the care, thought, and passion that animates many practicing scientists. This is not the whole story of what makes science possible, but I have never forgotten that it is a very important part.

Passionate thinking.[2] This ultimately is what supported this project throughout, and it found a home from the start to finish at the University of California, Santa Cruz (UCSC). When asked in my UCSC job interview why I would leave a private university on the East Coast to join a cash-strapped public university, I responded: because it is where my soul wants to be. I was young and naive. I also was lucky. UCSC has for decades fostered thought on the edge that has generated new modes of thinking and living. It helped a then new assistant professor foster a new domain of thought and action that in important ways formed this book: Science and Justice. I am grateful to the many mentors, students, colleagues, and administrative leaders who made its creation possible: Karen Barad, Mark Diekhans, Elaine Gan, Herman Gray, Donna Haraway, Lizzie Hare, David Haussler, Rusten Hogness, Cris Hughes, Zia Isola, Sheldon Kamieniecki, Martha Kenney, Paul Koch, Kristina Lyons, Colleen Massengale, Andrew Mathews, Jake Metcalf, Tyrus Miller, Ruth Müller, Ann Pace, Maria Puig de la Bellacasa, Natalie Purcell, Warren Sacks, Astrid Schrader, and the many graduate students who have taken part in the Science and Justice Training Program.[3] The support of other intellectual centers and research groups at UCSC also have been vital to Science and Justice and to the thought developed in this book, in particular the Center for Cultural Studies, the Institute for Arts and Sciences, and the History of Consciousness Department. The support of my colleagues in the genome sciences also has been vital. Thank you to David Haussler, Beth Shapiro, Ed Green, Rachel Harte, and Kate Rosenbloom for helping me to understand the latest developments in genomics. A special thanks to Mark Diekhans, genomics code writer extraordinaire, friend and ally in efforts to understand genomics and life at their fullest. Finally, but crucially, I would like to thank the staff, students, and colleagues in the Department of Sociology at UCSC for supporting a somewhat unconventional research trajectory, for unfailing grace and good humor under pressure, and for intellectual sustenance. Working with you has been a constant source of support and goodness in my life over this last decade. I want to particularly acknowledge the many undergraduate and graduate students who inspired me along the way, and who kept me attuned to developments near and far. Thank you especially to Catherine Weldon, Jocelyn Lee, and Michelle Peterson who were part of the dynamic *Postgenomic Condition Research Group*, and Hannah Finegold, Eric Alvarez, and Kate Darling who formed the fabulous *Just Data* team.

Understanding a global phenomenon like genomics requires extended stays in parts of the world far from home. I am grateful to the institutions and people who hosted and supported the research and writing of this book. In Edinburgh,

Scotland, the Genomics Forum and the University of Edinburgh hosted many rich and productive stays. Thank you to Gill Haddow, Jane Calvert, and Steve Sturdy for many a vigorous conversation and convivial meal. And to Steve, a special thank you for the hill walks, one through a very impressive amount of snow! I also would like to acknowledge the support of the Generation Scotland staff and researchers who graciously agreed to many interviews over the years, showed me around their labs and facilities, and even came to presentations of my research, offering critical commentary and feedback.

In the rest of the still United Kingdom, I am grateful to colleagues at the London School of Economics, King's College, the University of Sussex, the University of Cambridge, and the University of Warwick who over this last decade hosted me as a visiting scholar and provided critical feedback on many of the chapters of this book. Thank you in particular to Caroline Basset, Carlo Carduff, Carrie Friese, Amy Hinterberger, Kate O'Riordan, Bronwyn Parry, Barbara Prainsack, Nikolas Rose, and Ros Williams.

In the spring of 2013 I had the great fortune to spend two months on the gorgeous grounds of the Brocher Foundation on the banks of Lake Geneva. There, with beautiful sunsets and the gorgeous swans of the lake to inspire, the writing of this book began in earnest. I am grateful to Anyck Gérard, Marie Grosclaude, Elliot Guy, Philippe Pelet, and Roland Pellet for taking such good care of all of us who turn up on your shores in need of sustenance to carry forward the vision that Jacques and Lucette Brocher envisioned, one in which scientific and humanistic thought never part company. My thanks as well to the new colleagues I met during those two months—for the camaraderie and ping pong games—and especially to Marc Tennant for organizing us into a reading group that provided the motivation to finish first drafts of chapters 3 and 5.

In October 2014, I visited Berlin for the first of four visits during the final writing phase of this project. There I spent several very productive and rich months at the Max Planck Institute for the History of Science, first in the Histories of Knowledge about Human Variation Research Group and then the Artefacts, Action, and Knowledge Research Group. Many thanks to Veronika Lipphardt for the original invitation, for her group's critical engagement with my work, and the lively and ongoing intellectual exchanges. A special thank you also to Jenny Bangham and Sarah Blacker. Jenny helped me to understand the histories and shifting meanings of blood donation and data sharing in genetics and biomedicine. Both Jenny and Sarah became cherished companions in trying to understand how to live in the present with so many histories—painful and inspiring—pushing upon and guiding us. Finally, many thanks to Elisabeth Eppinger, who along with her then six-month-old daughter Agnes met me at the airport when I first arrived in Berlin, and who helped me over the last few years to create an unexpected and rich new home. First as hallmates at the Brocher Foundation, and then as cohosts of four East German cocktail parties, together we pushed at how we might redirect property regimes to help those in greatest need, and bridge ongoing political divides while Wir Mixen!

Finally, I would like to acknowledge the distributed worlds of science and technology studies, which supported a budding young scientist to explore science and knowledge in their deepest senses. The *Social Studies of Science* annual meetings as well as the formal and informal STS networks around the world make a project like *The Postgenomic Condition* possible. I would like to thank in particular my early mentors who continue to support me and these broader worlds: Sheila Jasanoff, Steve Hilgartner, and Evelynn Hammonds.

Support for travel to these places, as well as for the extended ethnographic fieldwork and historical research this book required, was made possible through the generous financial support of several institutions. The work was launched with the help of a grant from the National Science Foundation for my Paradoxes of Participation research (Award Number 0351475). It ended with the support of NSF Grant Number SES-1451684 that funded the *Just Data* workshop. The research that led up to and the discussions that unfolded at this workshop shaped my concluding thoughts. While I am grateful for the support of the National Science Foundation, of course the opinions, findings, conclusions, and recommendations expressed in this book are mine and do not necessarily reflect the views of the NSF.

Support for the research and writing of this book also came from a Distinguished Visiting Fellowship from the UK Economic and Social Science Research Council, and residencies at the Brocher Foundation and the Max Planck Institute for the History of Science. Finally, I would like to thank the UCSC Academic Senate Committee on Research and the Division of Social Sciences for supporting this work over its many years.

Ultimately, the writing of any book is supported through the collective energies of the many who take the time to respond to the work. For thoughtful and often formative engagement, I am thankful to the audiences at the University of Pennsylvania, the National Human Genome Research Institute, Northwestern University, Lancaster University, the University of Edinburgh, Stanford University, the University of Oxford, the New School, the University of Exeter, the University of Oslo, UC Berkeley, UC San Diego, the London School of Economics, King's College, Cornell University, UCLA, San Francisco State University, the University of Washington, Harvard University, the University of Cambridge, the University of Freiburg, the University of Copenhagen, Yale University, Ludwig Maximilian University of Munich, Kalamazoo College, and the University of Warwick. I want to particularly thank Joanna Radin and the graduate students at Yale University as well as Rosalind Williams and Amy Hinterberger at the University of Warwick for organizing workshops and extended engagements with my work and the broader domain of "science and justice."

Journals also provide critical support to the development of scholarship. I am grateful to the editors and blind reviewers at *Biosocieties*, the *Journal of Science Communication*, and *Personalized Medicine* where early versions of sections of chapters 4 and 5 first appeared.[4]

I am also enormously grateful to the friends and colleagues who read and provided critical feedback on one or many chapters of this book: Misha Angrist, Jenny Bangham, Sarah Blacker, Jason Bobe, Lisa Brooks, Bob Cook-Deegan, Angela Creager, Georgia Dunston, Troy Duster, Malia Fullerton, Gill Haddow, Julie Harris-Wai, Donna Haraway, Jean McEwan, Alondra Nelson, Kate O'Riordan, Skúli Sigurdsson, Ed Smith, Hallam Stevens, Nancy Stoller, and Steve Sturdy. Thank you for improving the thinking in this book, and for the care for and support of the work.

There are a few individuals who have been with me for this whole decade who provided critical intellectual sustenance and friendship throughout. To Kim TallBear, thank you for sharing the journey and the commitment to justice as a guiding force. To Alondra Nelson, our exchange of book chapters at a crucial point, and our mutual support, ensured that this book made it. Thank you. To Rebecca Herzig, your early and ongoing belief in the way I walk through and see this world is foundational to the possibility of this one life of mine. Thank you for reminding me to stay on the path, always. To Donna Haraway, thank you for the passionate thought, for helping me to see what I bring to our worlds, and for the grounded, sustaining friendship. And to Kate O'Riordan, who I am sure heard me discuss the ideas in every single one of these chapters more times than I would like to admit, words fail. Thank you for seeing, supporting and encouraging me as a thinker and person over this decade.

My gratitude also to those who made sure I stayed sane as I finished this decade-long project. Thank you to Colleen Massengale, Kate Darling, Joe Klett, Nancy Stoller, Becca Gutman, Sarah Blacker, and Jenny Bangham for cheering me on and for the support in the final months. And to Brixie, the bike that took me from London to Berlin in the final summer of this project, thank you for the journey and for renewing my wonder and care for the world. I will see you soon! Finally, a special thank you to Hannah Finegold who tirelessly and expertly worked with me in the final weeks to ensure that all endnotes, references, and images were in order.

I end by thanking the University of Chicago Press. Over the last few years I have developed a renewed and deepened appreciation for the practices of writing and forming books. In a world that today feels filled with a threatening thoughtlessness, about which Arendt worried, the materialization of reflective and passionate thought on the printed and digital page provides me with hope for collective life. And not just bare life, but full life. Life worth living. It is this commitment to the book in its deepest sense that I have found at the University of Chicago. To my editor, Karen Darling, thank you for being the quiet, powerful force that moved these words into the light of day, but who also helped the author to see the world outside, too. For all the advice and support on how to craft words and arguments that could communicate broadly, and for the encouragement to cast out on my bike, to never forget life beyond the screen, I am deeply grateful. Finally, I am very grateful to Evan White for helping me with image permissions, to Dawn Hall for careful attention to the final text, and for three blind reviewers whose astute and careful readings made this a much better book.

—San Francisco, November 21, 2016

NOTES

CHAPTER ONE

1 Thurston, "Why I'm Joining Helix as CEO." In 2015, Helix received $100 million from the sequencing giant Illumina and investment firms Warburg Pincus and Sutter Hill Venture.

2 As the *MIT Technology Review* put it, Helix seeks to be "an on-line store for information about your genes." Regaldo, "Apple Has Plans for Your DNA."

3 Ibid. See also "About the Precision Medicine Initiative®," National Institutes of Health, accessed August 17, 2016, https://www.nih.gov/precision-medicine-initiative-cohort-program.

4 Joyner, Paneth, and Ioannidis, "What Happens When Underperforming Ideas Become Entrenched?"

5 Pollack, "Awaiting the Genome Payoff."

6 Clinton, "Statement Transcribed on Decoding of Genome," D8.

7 "It is now conceivable that our children's children will know the term *cancer* only as a constellation of stars," stated US President Bill Clinton in the White House celebration of the draft of the human genome. Clinton, "Statement Transcribed on Decoding of Genome," D8.

8 Clinton proclaimed: "One of the great truths to emerge from this triumphant expedition inside the human genome is that in genetic terms, all human beings, regardless of race, are more than 99.9 percent the same. Clinton, "Statement Transcribed on Decoding of Genome," D8.

9 For a rich, historical exploration of how the biological problem of DNA-based protein synthesis came to be represented as a problem of information, and then of writing "the book of life," see Kay, *Who Wrote the Book of Life? A History of the Genetic Code.*

10 Notable exceptions include Fortun, *Promising Genomics: Iceland and DeCODE Genetics in a World of Speculation*; O'Riordan, *Genome Incorporated.*

11 The effort to sequence the human genome captured the public attention when Craig Venter and his privately backed company Celera challenged Francis Collins and NIH's effort to sequence the human genome. As I explore in the next chapter, popular accounts framed the effort as a "war" of scientific titans racing to sequence the human genome in order to either "save the world," save "our common heritage," or save public science. See Sulston and Ferry, *Common Thread: A Story of Science, Politics, Ethics, and the Human Genome*; Shreeve, *Genome War: How Craig Venter Tried to Capture the Code of Life and Save the World.*

12 As we will see, the defining story of the effort to sequence the human genome was the so-called race between the private- and publicly funded effort to sequence the human genome and the effort to save the human genome from enclosure by private companies and patents. Much ink has been spilled debating whether the scientific leader of the private effort to sequence the human genome, Craig Venter, was an egomaniac. Cook-Deegan, *Gene Wars: Science, Politics, and the Human Genome*; Shreeve, *Genome War: How Craig Venter Tried to Capture the Code of Life and Save the World*; Sulston and Ferry, *Common Thread: A Story of Science, Politics, Ethics, and the Human Genome*.

13 Baker, "Housing Crash Recession: How Did We Get Here?"

14 The building of a wall between the United States and Mexico—which began as a fence in the 1990s—is part of a global rise in border walls associated with the rise of the security state. See Jones, "Death in the Sands."

15 While trust in institutions has been dropping since the 1960s, prior to the financial crisis, most people still believed that large established institutions could know and ensure the value of their investments. After the crisis, deepening worldwide distrust in dominant institutions led many to ask what, if anything, still had worth, and who could answer this critical question. Fournier and Quinton, "How Americans Lost Trust in Our Greatest Institutions."

16 Freedland, "Post-Truth Politicians Such as Donald Trump and Boris Johnson Are No Joke."

17 Swaine et al., "Young Black Men Killed by US Police at Highest Rate in Year of 1,134 Deaths"; Williams and Blinder, "Baton Rouge Attack Deepens Anguish for Police: 'We've Seen Nothing Like This.'"

18 Sheila Jasanoff diagnosed this problem in the life sciences. The new life sciences, she argued, move beyond bioethical concerns about informed consent and privacy to constitutional questions about who citizens are and what makes up their fundamental rights. Her book develops this strand of scholarship. Jasanoff, *Reframing Rights: Bioconstitutionalism in the Genetic Age*.

19 On the need to step away from already-formed theories and to attend to how elements in the world congeal to form the things that garner care and concern, see also Arendt, *Origins of Totalitarianism*. Arendt's attention to elements is explored in Disch, "More Truth Than Fact: Storytelling as Critical Understanding in the Writings of Hannah Arendt," 665–94.

20 For the problem of limited care and concern, see Puig de la Bellacasa, "Matters of Care in Technoscience: Assembling Neglected Things," 85–106. For an understanding of truths and facts as not already-forged things that the law can enroll to administer justice, but as things that the values and orientations of the law help to create and stabilize, see Jasanoff, *Science at the Bar: Law, Science, and Technology in America*.

21 Paul, *Controlling Human Heredity: 1865 to the Present*; Kevles, *In the Name of Eugenics: Genetics and the Uses of Human Heredity*.

22 Catherine Bliss brings an account of these efforts up to the present in her book *Race Decoded: The Genomic Fight for Social Justice*.
23 UNESCO, *What Is Race?*
24 The *Annals of Eugenics* became the *Annals of Human Genetics*. The Galton Eugenics Professorship became the Galton Professorship of Human Genetics. Kevles, *In the Name of Eugenics*, 252.
25 By the 1960s, many human geneticists favored molecular analyses of blood and proteins over measurements of skulls and bodies. Silverman, "The Blood Group 'Fad,'" 11–27.
26 Cavalli-Sforza et al., "Call for a Worldwide Survey of Human Genetic Diversity: A Vanishing Opportunity for the Human Genome Project," 490–91.
27 Reardon, *Race to the Finish: Identity and Governance in an Age of Genomics*.
28 Friedlander, "Genes, People, and Property: Furor Erupts over Genetic Research on Indigenous Groups," 22–24; "Patents, Indigenous, and Human Genetic Diversity."
29 Herrnstein and Murray, *Bell Curve: Intelligence and Class Structure in American Life*, 77.
30 See Schultze, Dickens, and Kane, "Does the Bell Curve Ring True? A Closer Look at a Grim Portrait of American Society."
31 Author's field notes, April 29, 2015. Donohue, "Beyond Ethics: The Scientific and Technological Development of the International HapMap Project 1998 to 2005."
32 As a program officer at NHGRI explained to me: "We really need to put human back into the sequencing program. This is the National Institutes of Health. We do need, both for medical reasons of doing stuff that's useful to medicine, and answering to Congress, we need to say that we've got a human component to the sequencing program." Interview with author, May 30, 2005.
33 International HapMap Consortium, "Integrating Ethics and Science in the International HapMap Project," 5.
34 While Catherine Bliss documents genome scientists' efforts to construct a new antiracist "science of race," the period she investigates follows that of the HapMap. In the early 2000s, genome scientists still avoided the use of the term *race*, and many still denied it had biological meaning. Bliss, *Race Decoded: The Genomic Fight for Social Justice*; Angier, "Do Races Differ? Not Really, Genes Show."
35 International HapMap Consortium, "Integrating Ethics and Science in the International HapMap Project," 5.
36 Ibid.
37 Reardon, "The Democratic, Anti-Racist Genome? Technoscience at the Limits of Liberalism," 25–47; Horton, *Race and the Making of American Liberalism*; Starr, *Freedom's Power: The True Force of Liberalism*.
38 Avey, "Reminiscing."

39 Church, "The Personal Genome Project."

40 Ball et al., "Harvard Personal Genome Project: Lessons from Participatory Public Research," 10; Bobe, "Open Humans Network: Find Equitable Research Studies and Donate Your Personal Data for Public Benefit."

41 Clinton, "Statement Transcribed on Decoding of Genome," D8.

42 Locke, *Two Treatises of Government*; Starr, *Freedom's Power: The True Force of Liberalism*.

43 Interview with Francis Collins, September 11, 2013.

44 Scherer, "Secret Sharers: A Close Look at Privacy, Surveillance, and Hacktivism," 20–27.

45 See "I Want My Genome," *Genome Web*, accessed March 27, 2014, http://www.genomeweb.com/blog/i-want-my-genome.

46 Cohen, "Egyptians Were Unplugged, and Uncowed."

47 Today, not just the value of genomes, but that of social media, is under question. Notably, many now question the centrality of Facebook and Twitter to the Arab Spring. Scholars and activists argue that calling this uprising a social media revolution marginalizes the role of those who did not have the resources to access social media, ignores the role of social media in amplifying polarization, and fails to face the limits of a platform owned by private corporations. See Shearlaw, "Egypt Five Years On: Was It Ever a 'Social Media Revolution'?"

48 Castells, *The Rise of the Network Society*; Dean, *Publicity's Secret: How Technoculture Capitalizes on Democracy*.

49 See "Genomics Gets Personal: Property, Persons, Privacy."

50 Bowker and Star, *Sorting Things Out: Classification and Its Consequences*.

51 Skloot, *The Immortal Life of Henrietta Lacks*.

52 Axelrod, "The Immortal Henrietta Lacks."

53 Gilroy, *Against Race: Imagining Political Culture beyond the Color Line*; Landecker, "Between Beneficence and Chattel: The Human Biological in Law and Science"; Wald, "Cells, Genes, and Stories: Hela's Journey from Labs to Literature."

54 Goldberg, *Threat of Race: Reflections on Racial Neoliberalsim*; Jones, *Bad Blood: The Tuskegee Syphilis Experiment*; Reverby, *Examining Tuskegee: The Infamous Syphilis Study and Its Legacy*.

55 Sunder Rajan, *Biocapital: The Constitution of Postgenomic Life*.

56 Skloot, "The Immortal Life of Henrietta Lacks, the Sequel."

57 Brainard, "HeLa-Cious Coverage: Media Overlook Ethical Angles of Henrietta Lacks Story."

58 Goldberg, *The Threat of Race: Reflections on Racial Neoliberalsim*.

59 For an analysis of genome scientists' self-proclaimed efforts to advance the cause of social justice and antiracism, see Bliss, *Race Decoded: The Genomic Fight for Social Justice*.

60 For a video of the panel discussion, see "Genomics Gets Personal: Property, Persons, Privacy."

61 As Sunder Rajan makes clear, and as the stories in this book illustrate, the form capitalism takes in an era of biocapital—the source of value and the basis of exchange—are far from clear and settled. See Sunder Rajan, *Biocapital: The Constitution of Post-Genomic Life*, 7.

62 See MeForYou.org. Accessed December 10, 2013.

63 Desmond-Hellman, "Toward Precision Medicine: A New Social Contract?"

64 See http://www.meforyou.org/#do-it, accessed June 4, 2013. This website launched in May 2013 as the public relations component of the OME Summit, what UCSF described as "a collective effort by nearly 150 select participants to harness the power of personal and cumulative health data to fundamentally shift the practice of medicine," http://www.ucsf.edu/news/2013/01/13455/what-ome, accessed June 4, 2013.

65 Sandel, *Justice: What's the Right Thing to Do.*

66 Ibid., 18.

67 Etzioni, "Justice: What's the Right Thing to Do? Review."

68 The Boston Women's Health Book Collective, *Our Bodies, Ourselves: A New Edition for a New Era.*

69 Locke, *Two Treatises of Government.*

70 Liptak, "Justices Allow DNA Collection after an Arrest," A1.

71 See, for example, MeForYou.org, Weconsent.us, and the new Global Alliance to Enable Responsible Sharing of Genomic and Clinical Data described at http://www.broadinstitute.org/files/news/pdfs/GAWhitePaperJune3.pdf, accessed June 7, 2013.

72 For my account of this experience, see http://www.sfgate.com/opinion/article/Should-patients-understand-that-they-are-research-4321242.php, accessed October 14, 2016. For UCSF's response, see http://www.sfgate.com/opinion/openforum/article/UC-advances-medical-research-through-trust-4321228.php, accessed October 14, 2016. For the *San Francisco Chronicle*'s accompanying editorial, see http://www.sfgate.com/opin ion/editorials/article/Biological-material-laws-need-refining-4321235.php, accessed October 14, 2016.

73 See, for example, Picciano, "Why Big Data Is the New Natural Resource." For an analysis of how life was drawn into economic production, see Cooper, *Life as Surplus: Biotechnology and Capitalism in the Neoliberal Era.*

74 Spiegel. "Spiegel interview with Craig Venter: 'We Have Learned Nothing from the Genome.'"

75 For an excellent analysis of how a simple determinist notion of the action of genomes in the HGP era yielded to an emphasis on uncertainties and complexities in the postgenomic era, see Richardson and Stevens, *Postgenomics: Perspectives on Biology after the Genome.*

76 Recent studies also call into question the value of investments in biomedical research. See, for example, Bowen and Casadevall, "Increasing Disparities between Resource Inputs and Outcomes, as Measured by Certain Health Deliverables, in Biomedical Research."

77 GenomeWeb, "NIH to Pump up to $96m into New Big Data Centers."

78 Green et al., "A Draft Sequence of the Neandertal Genome," 710–22; Lorenzen et al., "Species-Specific Responses of Late Quaternary Megafauna to Climate and Humans," 359–64; Shapiro and Hofreiter, "A Paleogenomic Perspective on Evolution and Gene Function: New Insights from Ancient DNA."

79 "Using DNA Matching to Crack Down on Dog Droppings"; Nelson, *Social Life of DNA: Race, Reparations, and Reconciliation after the Genome.*

80 Bolnick et al., "The Science and Business of Genetic Ancestry Testing," 399–400; Darling et al., "Enacting the Molecular Imperative: How Gene-Environment Interaction Research Links Bodies and Environments in the Post-Genomic Age"; Nelson, *Social Life of DNA: Race, Reparations, and Reconciliation after the Genome.*

81 Hayden, "Genome Sequencing Stumbles towards the Clinic."

82 Brockman, *Digerati: Encounters with the Cyber Elite.* Linked to these questions about whether benefits are hyped are questions about whether the harms are as well. Popularized in the 1997 film *GATTACA*, fears of discrimination and the creation of a genetic underclass are widespread. However, many legal and social science scholars, as well as biomedical scientists, question the evidential basis of these fears. See Wertz, "Genetic Discrimination—an Overblown Fear?," 496.

83 O'Riordan, *The Genome Incorporated*; Stiglitz, *The Price of Inequality: How Today's Divided Society Endangers Our Future.*

84 Kanter and Sengupta, "Europe Continues Wrestling with Online Privacy Rules."

85 Huet, "Google Bus Blocked by Dancing Protesters in S.F. Mission District."

86 Knight, "In Growth of Wealth Gap, We're No. 1."

87 The danger of writing a book about genomics is that it may promote genomic exceptionalism. To be clear, I do not think that genomics raises special problems or concerns. Today it is not exceptional, but rather mainstream. It embodies the promises and dangers of the day.

88 On bioconstitutionalism, see Jasanoff, *Reframing Rights: Bioconstitutionalism in the Genetic Age.* For theorizations of the centrality of the life sciences to the contemporary constitution of societies, see Rose, *Politics of Life Itself: Biomedicine, Power, and Subjectivity in the Twenty-First Century.*

89 A number of recent books by cultural theorists describe the fraying of hope in post–World War II understandings of how to live a good life. See, for example, Berlant, *Cruel Optimism.*

90 Arendt, *Human Condition*; Lyotard, *Postmodern Condition: A Report on Knowledge*.

91 Many hoped that investment in science and technology would boost war-torn economies and provide objective grounds on which to rebuild trust in governance. See, for example, Polanyi, "Republic of Science: Its Political and Economic Theory," 54–72.

92 The quote continues: "In this case, it would be as though our brain, which constitutes the physical, material condition of our thoughts, were unable to follow what we do, so that from now on we would indeed need artificial machines to do our thinking and speaking. . . . Wherever the relevance of speech is at stake, matters become political by definition, for speech is what makes man a political being. If we would follow the advice, so frequently urged upon us, to adjust our cultural attitudes to the present status of scientific achievement, we would in all earnest adopt a way of life in which speech is no longer meaningful. For the sciences today have been forced to adopt a 'language' of mathematical symbols, though it was originally meant only as an abbreviation for spoken statements, now contains statements that in no way can be translated back into speech." Arendt, *Human Condition*, 3–4.

93 Hunter, Altshuler, and Rader, "From Darwin's Finches to Canaries in the Coal Mine—Mining the Genome for New Biology," 2760–63.

94 Arendt, *Human Condition*, 178–79.

95 Ibid., 5.

96 Lyotard, *Postmodern Condition: A Report on Knowledge*, 3.

97 Ibid., 8.

98 Stevens, *Life out of Sequence: A Data-Driven History of Bioinformatics*.

99 Venter, *A Life Decoded: My Genome, My Life*.

100 Ibid., 232.

101 Sulston and Ferry, *Common Thread: A Story of Science, Politics, Ethics, and the Human Genome*, 222.

102 Angrist, *Here Is a Human Being: At the Dawn of Personal Genomics*, 78–80.

103 Hayden, "Technology: The $1,000 Genome."

104 See "Illumina Enterprise Value," YCharts, accessed October 16, 2016, https://ycharts.com/companies/ILMN/enterprise_value.

105 Hayden, "Technology: The $1,000 Genome"; Regaldo, "Illumina Says 228,000 Human Genomes Will Be Sequenced This Year"; Timmerman, "DNA Sequencing Market Will Exceed $20 Billion, Says Illumina CEO Jay Flatley."

106 See, for example, Nelkin and Lindee, *DNA Mystique: The Gene as a Cultural Icon*.

107 See "Nobel Winner in 'Racist' Claim Row," CNN, accessed June 10, 2013, http://www.cnn.com/2007/TECH/science/10/8/science.race/index.html.

108 Bliss, *Race Decoded: The Genomic Fight for Social Justice*.

109 Owens and King, "Genomic Views of Human History," 451–53; Sorensen, "Race Gene Does Not Exist, Say Scientists"; Wells, *Journey of Man: A Genetic Odyssey.*

110 Sulston and Ferry, *Common Thread: A Story of Science, Politics, Ethics, and the Human Genome.*

111 Avey, "Reminiscing."

112 Rabinow, *French DNA: Trouble in Purgatory.*

113 The move to automation—from human to machines—did not concern Arendt because it might produce "the much deplored mechanization and artificialization of natural life," but because machines intensified life's tendency to metabolize, or consume. The result: "Machines . . . would not change, only make more deadly, life's chief character with respect to the world, which is to wear down durability." Thus they would create "the grave danger that eventually no object of the world will be safe from consumption and annihilation from consumption." Arendt, *Human Condition*, 132–33.

114 Ibid.

115 Latour and Weibel, *Making Things Public: Atmospheres of Democracy.*

116 Puig de la Bellacasa, "Matters of Care in Technoscience: Assembling Neglected Things," 85–106.

117 For a list of the animals for which we now have published genomes, see "List of Sequenced Animal Genomes," Wikipedia, accessed March 26, 2016, https://en.wikipedia.org/wiki/List_of_sequenced_animal_genomes#Urochordates.

118 Ramos et al., "Characterizing Genetic Variants for Clinical Action," 93–104.

119 Arendt, *Human Condition*, 182. For an excellent analysis of Arendt's conception of storytelling as a form of "situated impartiality," see Disch, "More Truth Than Fact: Storytelling as Critical Understanding in the Writings of Hannah Arendt," 665–94.

120 Arendt, *Human Condition*, 6.

121 Haraway, *When Species Meet.*

122 Teilhard de Chardin, *Future of Man.*

123 Angier, "Scientist at Work: Mary-Claire King; Quest for Genes and Lost Children."

124 In the early 1990s, the National Science Foundation invested significant resources into developing this new field. One goal was to understand fundamental questions about how the nature of knowledge and society might be changing as governments and corporations invested at unprecedented levels in the natural and physical sciences following World War II.

125 Reardon, *Race to the Finish: Identity and Governance in an Age of Genomics.*

126 Cavalli-Sforza, "Race Differences: Genetic Evidence"; Friedlander, "Genes, People, and Property."

127 Some Native peoples, for example, feared that genetics would be used to challenge their first settler status, and thus claims to lands and tribal membership. TallBear, "Narratives of Race and Indigeneity in the Genographic Project."

128 This book is based on nearly one hundred semistructured in-depth interviews conducted between 2000 and 2015, as well as fieldwork at sites in the United States, Mexico, England, and Scotland. Fieldwork and interviews conducted at each site are described in each chapter. The interviews and fieldwork for this book were reviewed by UC Santa Cruz's Institutional Review Board, but deemed exempt. However, I chose to use informed consent forms so that the goals and purposes of the interviews were clear to all who participated. Except in cases of public figures, and in the last main chapter on openness, I chose to keep identities of informants anonymous. I did so to help attune the reader to the public character of the problems and dilemmas described, and not to personal opinions.

129 As Donna Haraway, perhaps the most influential theorist of our times on issues of science and society, explains: "Why tell stories like this, when there are only more and more openings and no bottom lines? Because there are quite definite response-abilities that are strengthened in such stories." Haraway, "Awash in Urine: Des and Premarin® in Multispecies Response-Ability," 301–16.

CHAPTER TWO

1 Greene, "Seven Decades Ago, A New, Enormous Kind of Explosion."

2 Moore, *Disrupting Science: Social Movements, American Scientists, and the Politics of the Military, 1945–1975.*

3 Rob Nixon recently coined the term *slow violence* to describe these kinds of destructive forces that operate on different temporal scales, often imperceptible to unaided human senses. Genomics, of course, is anything but slow. Instead, as Mike Fortun argues, it is defined by its speed. However, the problem of justice it poses is similar to the one that Nixon describes: it operates on scales that defy representation and thus decisive judgment. Fortun, "Human Genome Project and the Acceleration of Biotechnology," 182–201; Nixon, *Slow Violence and Environmentalism of the Poor.*

4 Seventy years later, an answer to this question continues to elude. For a history of efforts to study the effects of radiation, and the problems of knowledge it poses, see Lindee, "Survivors and Scientists: Hiroshima, Fukushima, and the Radiation Effects Research Foundation, 1975–2014," 184–209.

5 Baldwin, *Evidence of Things Not Seen.*

6 Cook-Deegan, "Alta Summit, December 1984," 661–63.

7 Bird, *American Prometheus: The Triumph and Tragedy of J. Robert Oppenheimer*, 6.

8 Sulston and Ferry, *Common Thread: A Story of Science, Politics, Ethics, and the Human Genome*, 8.
9 For a discussion of the origin of this metaphor, see Keller, Secrets of Life, Secrets of Death: Essays on Language, Gender, and Science.
10 Ibid., xiii.
11 The term *moral economy* was coined by E. P. Thompson in the 1960s to describe the norms and practices that regulated exchange during the eighteenth-century bread riots. It was imported to the history of science by Robert Kohler to describe *Drosophila* geneticists' practices of exchange. I will explore its roles as a normative and descriptive analytic tool later in this chapter.
12 This chapter is based on four in-depth interviews I conducted with key architects of the Human Genome Project (HGP) between 2010 and 2015, as well as oral histories with key players in the HGP conducted both by the Duke Center for Public Genomics and Cold Spring Harbor Laboratory. It is also informed by a major meeting the UCSC Science and Justice Research Center convened in November 2015 on the history of the Bermuda Principles that gathered key genome scientists present at Bermuda with historians of genomics. See "The Genomic Open: Then and Now," Science and Justice Research Center, accessed October 14, 2016, http://scijust.ucsc.edu/november-18-2015-the-genomic-open-then-and-now/.
13 For a definition of informatic capitalism as I use it, see Franklin, "Virality, Informatics, and Critique; or, Can There Be Such a Thing as Radical Computation?," 153–70. Key components include the rise of information as a dominant commodity form, the coproduction of markets and informatics and their reworking of both labor and knowledge, and the incorporation of the goals of "democracy" and inclusion into these new modes of production and science.
14 I focus primarily on the accounts of the main protagonist and antagonist: John Sulston and Craig Venter.
15 Lincoln, "Gettysburg Address."
16 Sulston and Ferry, *Common Thread: A Story of Science, Politics, Ethics, and the Human Genome*, 8.
17 Collins, "Contemplating the End of the Beginning," 641–43.
18 Guthrie, *This Land Is Your Land: The Asch Recordings Volume 1*, track 14.
19 Genome Research, 2001.
20 Sulston and Ferry, *Common Thread: A Story of Science, Politics, Ethics, and the Human Genome*, 168–69.
21 Genome Research, 2001.
22 The contrasts are striking. Craig Venter had a yacht and hung out with men in suits. The public effort handed out gothic T-shirts from Rasputins in Berkeley to motivate its people's army. Ashburner, *Won for All: How the* Drosophila *Genome Was Sequenced*.

23 See Gumz, "UCSC Goes Bald to Back Childhood Cancer Research."
24 Ashburner, *Won for All: How the* Drosophila *Genome Was Sequenced*, 77.
25 Merton, "Notes on Science and Democracy," 115–26, 121–22.
26 U.S. v. American Bell Telephone Co., cited in Merton, "Notes on Science and Democracy," 123.
27 Vermeir and Margocsy, "States of Secrecy: An Introduction," 1–12, 2.
28 Shapin, *Scientific Life: A Moral History of a Late Modern Vocation*, 174.
29 The idea that openness is a central norm of science dates back at least as far as the seventeenth century. Francis Bacon, for example, argued that an open inquiry of nature counteracted superstition and was essential to creating reliable knowledge. The Royal Society formally adopted this principle of openness. However, critical studies of the norm of openness reveal their implementation of the norm is always partial. See Eamon, *Science and the Secrets of Nature: Books of Secrets in Medieval and Early Modern Culture*, 319.
30 Hilgartner, "Selective Flows of Knowledge in Technoscientific Interaction: Information Control in Genome Research," 267–80.
31 The term *technoscience* was coined by the science studies scholar Bruno Latour in order to reject the convention of distinguishing between science and technology. Along with the historian of science Steven Shapin, I argue that in previous historical periods the distinction is analytically useful as it points to the different institutions and cultures that defined and supported scientists and technologists. However, by the late twentieth century technological innovation moved to the heart of science and these distinctions largely disappeared. Within the natural sciences, genomics played a central role in this transformation. See Shapin, *Scientific Life*, 8, and Latour, *Science in Action*, 259.
32 Diamond v. Chakrabarty, 447 U.S. 303 (1980).
33 Sulston's job within the lab was to map the cell lineage of the worm *C. elegans*. Brenner chose the worm as a model organism because it had enough biological complexity to express behavioral traits but was simple and grew quickly enough to make genetic crossing and analysis possible. Ankeny, "Natural History of *Caenorhabditis elegans* Research," 474–79.
34 Sulston and Ferry, *Common Thread: A Story of Science, Politics, Ethics, and the Human Genome*, 21.
35 Kohler, *Lords of the Fly*: Drosophila *Genetics and the Experimental Life.*
36 For Sulston's account of the importance of sharing in the work of genetic mapping projects, see Sulston and Ferry, *Common Thread: A Story of Science, Politics, Ethics, and the Human Genome*, 70.
37 Kohler, *Lords of the Fly*: Drosophila *Genetics and the Experimental Life*, 133.
38 Dr. John Sulston, interview by Lina Lu.
39 Sulston and Ferry, *Common Thread: A Story of Science, Politics, Ethics, and the Human Genome*, 166.

40 Sulston explained that participants of the first Bermuda meeting arrived "with claims on pieces of paper announcing their intention to sequence a particular region, and during the course of the meeting any competing claims were sorted out." The results of these negotiations were recorded on a website called the Human Sequencing and Mapping Index, which was managed by the Human Genome Organization (HUGO). See Sulston and Ferry, *Common Thread: A Story of Science, Politics, Ethics, and the Human Genome*, 166.
41 For an explanation of how fly geneticists attempted to craft notions of property that fostered communal ownership, see Kelty, "This Is Not an Article."
42 As Robert Kohler explains, a moral economy refers to the "nonmonetary obligations and rights of access to the necessities of life." See Kohler, *Lords of the Fly*: Drosophila *Genetics and the Experimental Life*, 12. The term was coined by E. P. Thompson to describe the rules, customs, and obligations that regulated the market for basic foodstuffs. Thompson, *Making of the the English Working Class.*
43 Thompson, "Moral Economy Reviewed," 271.
44 Kohler recognizes the tension in the term *moral economy* between its descriptive and normative function; personal correspondence with author.
45 Sulston and Ferry, *Common Thread: A Story of Science, Politics, Ethics, and the Human Genome*, 94.
46 For an in-depth historical account of these on-the-ground realities at the time, see Jones, Ankeny, and Cook-Deegan, "Bermuda Triangle."
47 Interview with author, January 22, 2013.
48 As another leading genome scientist involved in sequencing the human genome explained to me in November of 2015, there was a part of the Bermuda openness policies that favored the wealthy over the less wealthy. Personal correspondence with author.
49 Interview with author, January 22, 2013.
50 Indeed, it was a critical issue in the ongoing struggle with Craig Venter. Venter proposed a "shotgun approach" to sequencing the human genome partly because he believed it would more quickly lead scientists to the portion of the genome that was meaningful—the genes—and would greatly speed up the sequencing of the entire genome. His opponents argued that this shotgun approach would produce serious gaps in the human genome sequence. It would sacrifice quality of data in favor of quantity and speed. This was not a trade-off many working on the public HGP were willing to make. See Sulston and Ferry, *Common Thread: A Story of Science, Politics, Ethics, and the Human Genome*, 119.
51 Johns, *Piracy: The Intellectual Property Wars from Gutenberg to Gates*, 237.
52 Interview with author, January 22, 2013.
53 Nor was the Bermuda meeting as idyllic as it sounds. Genome scientists convened in Bermuda in late February under gray skies and met in a room

with no windows in a pink hotel. See McElheny, *Drawing the Map of Life: Inside the Human Genome Project*, 117.

54 References to the prison door appear on pp. 13, 14, 86, 99, 207, and 266 of Sulston and Ferry's account.

55 Sulston and Ferry, *Common Thread: A Story of Science, Politics, Ethics, and the Human Genome*, 50.

56 Ibid., 177.

57 Ibid., 156, 205–6, 131.

58 Stevens, *Life out of Sequence: A Data-Driven History of Bioinformatics*, 104.

59 Sulston and Ferry, *Common Thread: A Story of Science, Politics, Ethics, and the Human Genome*, 189.

60 Ibid., 188–89.

61 Ibid., 203.

62 On the centrality of speed to the HGP, see Fortun, "Human Genome Project and the Acceleration of Biotechnology," 182–201.

63 Lyotard, *Postmodern Condition: A Report on Knowledge*, 5.

64 Steinberg argued that NIH did not have the resources to compete with Japan, Britain, and Germany. See Venter, *A Life Decoded: My Genome, My Life*, 158.

65 Sulston and Ferry, *Common Thread: A Story of Science, Politics, Ethics, and the Human Genome*, 247. The value returned when the US president's science advisor and Francis Collins clarified that the principle of openness did not prevent private companies from using the data.

66 Ibid.

67 This, of course, is Celera. Ibid., 220.

68 See, for example, Jodi Dean's analysis of the rise of communicative capitalism and the concomitant expansion of the infrastructures of the information society. Dean, *Publicity's Secret: How Technoculture Capitalizes on Democracy*, 3–4.

69 Sulston and Ferry, *Common Thread: A Story of Science, Politics, Ethics, and the Human Genome*, 238. See also Cukier, "Open Source Biotech: Can a Non-Proprietary Approach to Intellectual Property Work in the Life Sciences?"

70 Shorett, Rabinow, and Billings captured these changes at the time in their 2003 commentary in *Nature Biotechnology*. Shorett, Rabinow, and Billings, "Changing Norms of the Life Sciences," 123–25.

71 Enlightenment thinkers such as Jeremy Bentham argued that information and knowledge secured right judgment. Bentham, "Of Publicity," 310. For a contemporary description of the ongoing centrality of this view, see Dean, *Publicity's Secret: How Technoculture Capitalizes on Democracy*; Nunberg, "Data Deluge," 1, 10–11.

72 Sulston and Ferry, *Common Thread: A Story of Science, Politics, Ethics, and the Human Genome*, 8.

73 Ibid., 276–89.

74 House of Representatives, 2010.

75 Craig Venter, interview by Spiegel, "We Have Learned Nothing from the Genome.'"

76 Shannon quoted in Gleick, *The Information*, 221–22.

77 Terranova, *Network Culture: Politics of the Information Age*, 12.

78 Twitter does care about the meaning of tweets when they are deemed abusive. With the rise of the alt-right in the United States, and what many view as attacks on civil, rational discourse, attention to the question of the value and meaning of words will likely increase. See Andrews, "Twitter Suspends Prominent Alt-Right Accounts."

79 Dean, *Democracy and Other Neoliberal Fantasies: Communicative Capitalism and Left Politics.*

80 Terranova uses the term *milieu* to refer to the way in which information was not just the content of the communication but the "massless flows" in "which contemporary culture unfolds." Terranova, *Network Culture: Politics of the Information Age*, 8.

81 Venter, *A Life Decoded: My Genome, My Life*, 160, 190, 30, 79.

82 Collins, *Language of God: A Scientists Presents Evidence for Belief*, 90.

83 Ibid., 98.

84 Ibid., 100 (italics added).

85 See Terranova, *Network Culture: Politics of the Information Age*, 32–33, on the precision of digital over analog.

86 Venter, *A Life Decoded: My Genome, My Life*, 232.

87 Ibid., 100.

88 Ibid., 166.

89 In both the opening and final pages of *The Human Condition*, Arendt describes the rocketing from Earth enabled by the science and technology of her Sputnik age, and the problem of alienation and the evacuation of meaning she believes it created.

90 Venter, *A Life Decoded: My Genome, My Life*, 110.

91 Ibid., 200.

92 Ibid., 181.

93 Jameson, *Political Unconscious: Narrative as a Socially Symbolic Act*, 60–61.

94 Baudrillard, *Ecstasy of Communication*; Poster, *Mode of Information: Poststructuralism and Social Context.*

95 Nunberg, "Data Deluge," 1–10, 11.

96 Clarke, "Information Wants to Be Free . . ."

97 To power the computers needed to sequence the human genome, Venter reports that the local power company lacked adequate resources, and that Celera had to install a new transformer and power poles. Venter, *A Life Decoded: My Genome, My Life*, 251.

98 This quote comes from George Church's discussion with Charlie Rose on the *Charlie Rose* show on June 19, 2009. See http://www.charlierose.com/guest/view/6321, last accessed September 8, 2011.

99 It is hard to pinpoint an exact date for this. Was it when Clinton, Venter, and Blair announced the completion of the draft in 2001? Was it when the "final" draft was published in 2003? This lack of clarity points to the underlying problem: genome scientists built infrastructures suited to producing, and not interpreting, genomic information. It is not surprising then that the news coming out of genomics would continue to be improvements in the production of that information.

CHAPTER THREE

1 I put "diversity" in quotes as leaders at the National Human Genome Research Institute later disavowed it. They argued that they did not study "human genetic diversity," but "human genetic variation." Presumably they understood "variation" to be a more neutral, objective term than "diversity." Personal correspondence with NHGRI program officer, December 16, 1998.

2 Reardon, *Race to the Finish: Identity and Governance in an Age of Genomics.*

3 Friedlander, "Genes, People, and Property: Furor Erupts over Genetic Research on Indigenous Groups," 22–24; Swedlund, "Is There an Echo in Here? Historical Reflections on the Human Genome Diversity Project."

4 Author's field notes, Capturing the History of Genomics, Bethesda, Maryland, April 29, 2015.

5 Lewontin, "Apportionment of Human Diversity," 381–98.

6 Angier, "Do Races Differ? Not Really, Genes Show"; Fitzpatrick, "Biological Differences among Races Do Not Exist, WU Research Shows"; Hotz, "Scientists Say Race Has No Biological Basis," A1.

7 Gilroy, *Against Race: Imagining Political Culture beyond the Color Line.*

8 On the importance of the black civil rights movement to defining late twentieth-century notions of inclusion in the American polity, see Skrentny, *Minority Rights Revolution.*

9 Higginbotham, "African-American Women's History and the Metalanguage of Race," 251–74.

10 Specifically, I visited and interviewed researchers at Howard University, Tuskegee University, and North Carolina A&T.

11 This chapter is based on fieldwork and six in-depth interviews conducted with genetic counselors, geneticists, and bioethicists at Tuskegee University in spring 2000, follow-up interviews conducted in 2013 with key players in the effort to establish a genome sequencing center at Tuskegee, and five additional in-depth interviews with African American scholars who worked at the time at HBCUs that sought to establish genomics centers in the late 1990s and early 2000s.

12 Alondra Nelson provides an invaluable exploration of the ambiguities, contestations, and struggles over the meaning of genomics for racial justice in her most recent book, *Social Life of DNA: Race, Reparations, and Reconciliation after the Genome*.

13 For a discussion of the more general shift from "sciences of the actual" to speculative, anticipatory forms of biomedical practices, see Adams, Murphy, and Clarke, "Anticipation: Technoscience, Life, Affect, Temporality," 246–65.

14 Hotz, "Scientists Say Race Has No Biological Basis," Al.

15 Flint, "Don't Classify by Race, Urge Scientists," B1.

16 Du Bois, *Souls of Black Folk*.

17 According to the National Institutes of Diabetes and Digestive and Kidney Diseases, "African Americans" are six times more likely than "Caucasians" to develop hypertension-related kidney failure. See "High Blood Pressure and Kidney Disease," NIH, accessed July 1, 2013, http://kidney.niddk.nih.gov/kudiseases/pubs/highblood/.

18 Antigens are the body's way of sorting self from other, and play a crucial role in fighting off infections; it is essential that the body "recognizes" the transplanted organ as its own, and for this the HLA types must match.

19 For Dunston's own account of her research and career, see Dunston, "A Passion for the Science of the Human Genome," 4154–56.

20 See http://www.genomecenter.howard.edu/milestones.htm, accessed May 3, 2013. Multiparous women are those who have undergone multiple pregnancies. These multiple pregnancies increase HLA levels and thus ease the creation of reagents.

21 See http://www.genomecenter.howard.edu/milestones.htm, accessed May 13, 2013.

22 Cann, Stoneking, and Wilson, "Mitochondrial DNA and Human Evolution," 31–36.

23 As one African American biomedical scientist explained to me: "If . . . you wanted to be economically sound, and you simply asked the question where can I get the broadest measure of human diversity for the least amount of dollars, how can I do this? African Americans to me. Because they give you the depth in Africa, and the mixture gives you some of the more recent kinds of variations that may have come from the younger population. So, you have the old and the new, all kind of brought together." Interview with author, January 14, 1999.

24 Ibid.

25 Dausset et al., "Centre D'etude Du Polymorphisme Humain (Ceph): Collaborative Genetic Mapping of the Human Genome," 575–77.

26 At the same time that Georgia Dunston and her colleagues at the Howard University School of Medicine attempted to study genetic differences in

African American populations for the purposes of improving health outcomes, Michael Blakey, a professor of anthropology at Howard University, was innovating genetic techniques to identify the remains found in the African Burial Ground (ABG) in New York City. While researchers broadly embraced the use of genetic techniques in the ABG, the use of genetics to discern ancestry sparked many debates. I explore these debates in chapter 5. Blakey, "Beyond European Enlightenment: Toward a Critical and Humanistic Biology," 379–406.

27 Jackson, "Assessing the Human Genome Project: An African American and Bioanthropological Critique," 105–12.

28 Ibid., 99–101.

29 NIH's Research Centers in Minority Institutions (RCMI) Program supported the founding of this project at Howard. See http://grants.nih.gov/grants/guide/pa-files/PAR-11-132.html, accessed May 7, 2013.

30 For a history and sociological analysis of the significance of the turn to genetics to understand African American history, see Nelson, *Social Life of DNA*.

31 Haley, *Roots*.

32 At the time, historians were collecting and archiving slave ship records and making them publicly available for the first time. See http://www.slavevoyages.org/tast/about/history.faces, accessed June 12, 2013.

33 Interview with author, June 4, 1999.

34 Interview with author, January 14, 1999.

35 Interview with author, June 4, 1999.

36 Ibid. On the formulation and contestation of genomic representations as a form of reparations, see Nelson, *Social Life of DNA*.

37 Duster, *Backdoor to Eugenics*.

38 Duster identifies the shift from an industrial to postindustrial economy, a phenomenon that displaced many workers, as a major cause of this uptick (Duster, *Backdoor to Eugenics*, 6). For a further elaboration of the argument that structural changes in society, and not individual predispositions, explain patterns of inequality, see Fischer et al., *Inequality by Design: Cracking the Bell Curve Myth*.

39 Herrnstein and Murray, *Bell Curve: Intelligence and Class Structure in American Life*; Wailoo, *Dying in the City of the Blues: Sickle Cell Anemia and the Politics of Race and Health*.

40 Dunston, "G-RAP: A Model HBCU Genomic Research and Training Program," 106–7; Swedlund, "Is There an Echo in Here? Historical Reflections on the Human Genome Diversity Project"; Marks, "Letter: The Trouble with the HGDP," 243.

41 See Harry, "The Human Genome Diversity Project and Its Implications for Indigenous Peoples."

42 "Resolution by Indigenous Peoples," Indigenous Peoples Council on Biocolonialism, accessed March 5, 2002, http://www.ipcb.org/resolutions/index.htm.

43 Kahn, "Genetic Diversity Project Tries Again," 720–22; Lewin, "Genes from a Disappearing World," 25–29; Macilwain, "Diversity Project 'Does Not Merit Federal Funding,'" 774.

44 Cavalli-Sforza, "The Human Genome Diversity Project."

45 Swedlund, "Is There an Echo in Here? Historical Reflections on the Human Genome Diversity Project"; and Harry, "Patenting Life and Its Implications for Indigenous Peoples," 1–2.

46 Consider the explanation an NIH program officer offered of scientists' future willingness to adopt new ethical practices, such as community consultation: "I think for some people it was like 'whatever it takes, just so that we don't sort of fall into the Diversity Project trap.'" Interview with author, June 30, 2005.

47 Interview with author, January 12, 1999. As we will see, by 1997, NIH had changed its position and appointed a program director of the Genetic Variation program.

48 Herrnstein and Murray, *Bell Curve: Intelligence and Class Structure in American Life*, 405–10.

49 Duster, "Review Essay in Symposium on the Bell Curve," 158–61.

50 The field of behavior genetics has long been the source of controversy that unsurprisingly genome scientists sought to avoid. For an in-depth history of the field and analysis of the causes and functions of these controversies, see Panofsky, *Misbehaving Science: Controversy and the Development of Behavior Genetics*.

51 Indeed, claims linking genes to IQ and race soon appeared in prominent places such as the *New York Times*. Wade, "First Gene to Be Linked with High Intelligence Is Reported Found"; Wade, "Researchers Say Intelligence and Diseases May Be Linked in Ashkenazic Genes."

52 Interview with author, April 30, 2013.

53 Sankaran, "African American Genome Mappers Pledge to Carry on Despite Grant Rejection," 1.

54 Interview with author, June 4, 1999.

55 Interview with author, August 25, 1998.

56 Ibid. Samples for SNPs eventually came from the National Health and Nutrition Examination Survey (NHANES). See http://www.cdc.gov/nchs/nhanes.htm, accessed September 19, 2013. One hundred and twenty samples came from human beings the article described as "European Americans," 120 from "African Americans," sixty from "Mexican Americans," thirty from "Native Americans," and 120 from "Asian Americans." Not surprisingly, given the short but charged history of efforts to study human genetic variation,

this DNA Polymorphism Discovery Resource raised concerns. However, concerns arose not because NHGRI used American categories of race and ethnicity to structure the resource, but because it chose *not* to make available information about the racial and ethnic origins of samples, and explicitly forbade researchers from attempting to attribute racial and ethnic categories to the samples. This somewhat surprising decision, they explained, arose from ethical considerations. See Collins, Brooks, and Chakravarti, "A DNA Polymorphism Discovery Resource for Research on Human Genetic Variation," 1229–31.

57 Greenberg, "Special Oversight Groups to Add Protections for Population-Based Repository Samples," 745–47.

58 Epstein, *Inclusion: The Politics of Difference in Medical Research.*

59 Jones, *Bad Blood: The Tuskegee Syphilis Experiment*; Reverby, *Examining Tuskegee: The Infamous Syphilis Study and Its Legacy.*

60 Interview with author, June 7, 1999.

61 Epstein, *Inclusion: The Politics of Difference in Medical Research*, 75–76.

62 Ibid., 90.

63 Reverby, *Examining Tuskegee: The Infamous Syphilis Study and Its Legacy*, 218–19.

64 Ibid., 223.

65 Ibid., 222.

66 Ibid., 223.

67 Benjamin Payton, then president of Tuskegee University, envisioned the infusion of bioethics curriculum into all areas of the university. As the 2000 Continuation Proposal for the TU Bioethics Center explained, the center sought to "expose all freshman students at Tuskegee University to bioethical issues and thought processes," and "ensure that all Tuskegee graduates possess a working knowledge of bioethics issues that are appropriate to their career choices." Tuskegee University, "The Tuskegee University National Center for Bioethics in Research and Health Care: Continuation Application."

68 Smith, "Trickle-Down Genomics: Reforming 'Small Science' as We Know It," 19.

69 Cavalli-Sforza, "Race Differences: Genetic Evidence," 51–58.

70 The Belmont Report has guided US federal policy on research ethics since the 1970s. King, "Dilemma of Difference," 75–82.

71 According to the survey data collected from meeting participants, King's speech received an "overwhelming" positive response. Interview with author, May 1, 2000.

72 Human Genome News, "Ramping Up Production Sequencing," 3.

73 Specifically, he argued it would extend access of state-of-the-art genomic techniques to biologists with diverse interests, promote "broad participation" of

students in "advanced research," and ensure that "broad societal discussion of the future of genetics is informed by widespread familiarity with the technical activities of the Human Genome Project." Olson, "Genome Centers: What Is Their Role?," 17–26.

74 For a full text of the apology, see http://www.cdc.gov/tuskegee/clintonp.htm, accessed May 13, 2013.

75 Olson, "Genome Centers: What Is Their Role?," 26.

76 In Alabama, there are twelve counties in what is called the Black Belt. As Booker T. Washington explained in his autobiography in 1901: "So far as I can learn, the term was first used to designate a part of the country which was distinguished by the colour of the soil. The part of the country possessing this thick, dark, and naturally rich soil was, or course, part of the South where the Slaves were most profitable, and consequently they were taken there in the largest numbers. Later, and especially since the war, the term seems to be used in a political sense—that is, to designate the counties where the black people outnumber the white." Washington, *Up from Slavery*, 108.

77 The Duffy Antigen/Chemokine Receptor (or DARC) is the protein receptor for human malarial parasites *Plasmodium vivax* and *Plasmodium knowlesi*. Individuals with a particular form of the DARC gene known as *Duffy null* suppress this protein receptor and so are resistant to infection by *P. vivax* or *P. knowlesi*. In regions with high rates of malaria, human geneticists believed natural selection favored those who inherited it. In addition to *Duffy null*, they hypothesized genetic variants in the surrounding region also would be preserved. However, they did not know how far out the resulting linkage disequilibrium (LD), a measure of the association of genetic variants along the chromosome, extended. Leonid Kruglyak—then in Seattle at the Fred Hutchinson Cancer Center, just blocks from Olson and Smith at UW—had just proposed that the length of the region in which genetic variants would be linked, and thus inherited together, was only 3kb, or 3,000 bases. If this was true, then knowing one genetic variant would only lead you to be able to predict variants within 3,000 bases of this variant. This enabled too little prediction, and required too much actual genotyping, to make feasible a haplotype map—a map of genetic variation based on these powers of prediction. Given that at the time leaders at NHGRI expressed interest in funding a haplotype map, answering the question of how far LD extended was a high priority. Kruglyak, "Prospects for Whole-Genome Linkage Disequilibrium Mapping of Common Disease Genes," 139–44.

78 Reich et al., "Reduced Neutrophil Count in People of African Descent Is Due to a Regulatory Variant in the Duffy Antigen Receptor for Chemokines Gene."

79 Olson, "Genome Centers: What Is Their Role?" 26. Arising from the 1996 Plain Talk conference, the NHGRI established a genome research part-

nership program between HBCUs and Genome Centers. The partnership funded Smith's research stays at UW.

80 Interview with author, May 1, 2000.

81 See "Poverty," United States Census Bureau, accessed May 13, 2013, http://www.census.gov/hhes/www/poverty/data/census/2000/poppvstat00.html.

82 Interview with author, February 7, 2013.

83 Hamblin and Di Rienzo, "Detection of the Signature of Natural Selection in Humans: Evidence from the Duffy Blood Group Locus," 1669–79.

84 For a description of the project, see http://search.engrant.com/project/Tn596K/apolipoprotein_b_variation_and_coronary_heart_disease_risk_in_african_americans, accessed May 13, 2013.

85 By the mid-1990s, physician epidemiologist Richard Cooper and epidemiologist Charles Rotimi were writing papers suggesting that one could compare individuals from Africa to African Americans in the United States in order to understand whether diseases—such as hypertension—which are prevalent in African American communities are due to genes or environment. TU researchers imagined that the proposed Black Belt study would contribute to this line of research. Cooper et al., "The Prevalence of Hypertension in Seven Populations of West African Origin," 160–68.

86 The hope was to develop a "Framingham-like" study. The Framingham Heart Study began in 1948 with 5,209 subjects from Framingham, Massachusetts, who researchers studied over the course of the next several decades. In particular, researchers followed the development of heart disease in subjects and associated risk factors with each incident. Researchers at TU proposed a similar approach. However, this time instead of the middle-class white residents of Framingham, Massachusetts, TU planned to recruit African Americans from the Black Belt—2,000 in total. Interview with author, September 12, 2000.

87 NIH increasingly received requests from researchers to include more samples from individuals of different ethnic and racial backgrounds in their Human Genetic Cell Repository maintained at the Coriell Institute in Camden, New Jersey. Greenberg, "Special Oversight Groups to Add Protections for Population-Based Repository Samples," 745–47.

88 Indeed, this issue remains unresolved and plays a role in all the stories told in this book.

89 A 2011 Alabama State University study put illiteracy rates in Alabama at 25 percent. See "Macon County residents react to ASU study," WSFA News, accessed May 13, 2013, http://www.wsfa.com/story/15118931/macon-county-residents-react-to-asu-study?clienttype=printable.

90 Bonham et al., "Community-Based Dialogue: Engaging Communities of Color in the United States' Genetics Policy Conversation," 325–59.

91 Field notes of author, April 27, 2000.

92 Interview with author, May 1, 2000.

93 Interview with author, May 1, 2000. Troy Duster and Diane Beeson found a similar thing in their study of African American perspectives on genetic testing. One person they interviewed compared worrying about sickle cell disease to worrying about a broken window when the house is burning. See Beeson and Duster, "African-American Perspectives on Genetic Testing," 151–74.

94 Interview with author, April 28, 2000.

95 The poverty rate in Tuskegee in 2009 was 50.5 percent. See "Tuskegee, Alabama (AL) Poverty Rate Data," City-Data, accessed May 13, 2013, http://www.city-data.com/poverty/poverty-Tuskegee-Alabama.html.

96 Interview with author, April 28, 2000.

97 Ibid.

98 Ibid.

99 When I asked about the benefit of the research to the local community, I was told: "By participating in projects like ours they get used to . . . participating in a genetic project, especially if it is directed by the government, so when they go to the hospital, and somebody tells them 'Oh we will look at whether or not you will have sickle cell or your kids will have sickle cell,' I don't think it will alarm them. But we are not going to give that data away. This is just for us to know how to prepare for your health in the future. . . . Because at the moment they probably would not do it. But we have to change the mindset . . . by helping them to participate in our Project and projects like [ours]." Interview with author, May 1, 2000. It is understandable why researchers would make this argument, and why many potential subjects could be confused. The questions of science and justice it raises have not been adequately addressed, leading to discomfort. I will return to this point in the conclusion.

100 Interview with author, April 28, 2000.

101 Ibid. Infamously, Nobel Prize laureate Linus Pauling suggested that all people with sickle cell trait should have a tattoo placed on their forehead. Wailoo, *Dying in the City of the Blues: Sickle Cell Anemia and the Politics of Race and Health*.

102 Interview with author, June 1, 2005.

103 Interview with author, June 11, 2004.

104 However, it should be noted that leaders of the center also observed that the community did not recognize a need for bioethics: "Bioethics still confuses this community. They don't really know what it is. And I mean we can define it fifty million times, but it's not within the context of this world." Interview with author, April 28, 2000.

105 Author's field notes, May 1, 2000.

106 Francis Collins followed with the following eloquent words: "I'm happy that today, the only race we are talking about is the human race." Craig Venter

would concur: "In the five Celera genomes, there is no way to tell one ethnicity from another. Society and medicine treats us all as members of populations, where as individuals we are all unique, and population statistics do not apply." See http://www.ornl.gov/sci/techresources/Human_Genome/project/clinton2.shtml, accessed May 13, 2013.

107 King, *Interview with Patricia King: Belmont Oral History Project.*

108 These issues are brilliantly discussed in Susan Reverby's book *Examining Tuskegee: The Infamous Syphilis Study and Its Legacy.*

109 King, *Interview with Patricia King: Belmont Oral History Project.*

110 On the problem of future-oriented, promissory science, see Adams, "Anticipation: Technoscience, Life, Affect, Temporality," 246–65.

111 University of Mississippi, known to many as Ole Miss, has struggled for decades with problems of race. During the Kennedy administration in the early 1960s, Governor Ross Barnett resisted desegregation policies. Kennedy called out the National Guard to escort twenty-nine-year-old James Meredith so that he might be admitted to campus while Barnett gave a speech in which he defended Mississippi's "customs" and "heritage" to a Confederate-flag-waving overflow crowd in the university's stadium. In the 1980s, Dr. Richard Hutchinson developed a National Heart, Lung, and Blood Institute (NHLBI) funded longitudinal study of atherosclerosis at the University of Mississippi. In 1998, he successfully added a longitudinal study of heart disease in African Americans that would become known as the Jackson Heart Study. For information about the study, see http://jhs.jsums.edu/jhsinfo, accessed May 18, 2013. Researchers, such as UW's Deborah Nickerson, would end up collaborating with Hutchinson, not Smith, to obtain "African American" samples from the Black Belt.

CHAPTER FOUR

1 Not all Native Americans opposed participation in human genetic variation research. Bliss, *Race Decoded*, 60. Author's field notes, September 7, 2006. For an in-depth discussion of Native American concerns and engagement with DNA, see TallBear, *Native American DNA.*

2 Interview with author, July 27, 2006.

3 Interview with author, May 30, 2005.

4 "Background on Ethical and Sampling Issues Raised by the International HapMap Project," National Human Genome Research Institute, last modified May 22, 2012, https://www.genome.gov/10005337.

5 For an overview, see Wall and Pritchard, "Haplotype Blocks and Linkage Disequilibrium in the Human Genome," 587–97.

6 International HapMap Consortium, "Integrating Ethics and Science in the International HapMap Project," 472.

7 Foner, *Story of American Freedom.*

8 International HapMap Consortium, "Integrating Ethics and Science," 471–72. Scott, *Only Paradoxes to Offer: French Feminists and the Rights of Man.*

9 This chapter is based on fieldwork at some of the original meetings where the HapMap was discussed, as well as interviews with twenty-two of the main organizers and participants in the debates about the initiative conducted primarily between 2004 and 2007. I also conducted follow-up interviews with many of these individuals during this period, as well as in 2013 and 2014.

10 Collins quoted in Couzin, "New Mapping Project Splits the Community," 1391–93.

11 Interview with author, September 11, 2006.

12 Interview with author, November 10, 2005; interview with author, May 30, 2005.

13 Dembner, "Harvard-Affiliated Gene Studies in China Face Federal Inquiry"; Benjamin, "A Lab of Their Own: Genomic Sovereignty as Postcolonial Science Policy," 341–55.

14 Interview with author, May 30, 2005.

15 Interview with author, September 13, 2006.

16 Interview with author, May 30, 2005.

17 Interview with author, July 27, 2006.

18 Within these large majority populations, they sought to collect blood and DNA from communities with "robust" identities and strong political organizations. Interview with author, November 20, 2006.

19 International HapMap Consortium, "Integrating Ethics and Science in the International HapMap Project," 472.

20 Interview with author, May 30, 2005.

21 Interview with author, November 20, 2006; interview with author, September 13, 2006; Reddy, "Good Gifts for the Common Good: Blood and Bioethics in the Market of Genetic Research," 429–72.

22 *Bioconstitutionalism* is a term coined by Sheila Jasanoff to call attention to the way in which the life sciences now act as a terrain on which the fundamental rights and responsibilities of citizens are contested and forged. Paul Rabinow first coined the term *biosociality* in the late 1990s to describe the way that new social groups were forming around genetic identities. Jasanoff, "Reframing Rights: Bioconstitutionalism in the Genetic Age"; Rabinow, "Artificiality and Enlightenment: From Sociobiology to Biosociality."

23 Interview with author, August 14, 2007.

24 Executive Summary of the First Community Consultation on the Responsible Collection and Use of Samples for Genetic Research, available at https://www.nigms.nih.gov/News/reports/archivedreports2003-2000/Pages/community_consultation.aspx, accessed December 11, 2015.

25 Interview with author, October 27, 2006.

26 Ibid.

27 Ibid.

28 All descriptions and quotes from this section derive from the first version of the "Policy for the Responsible Collection, Storage, and Research Use of Samples from Identified Populations for the NIGMS Human Genetic Cell Repository" published at http://locus.umdmj.edu/nigms/comm/submit/collpolicy.html, accessed January 22, 2002. The most recent version of the policy can be read at http://www.nigms.nih.gov/Research/SpecificAreas/HGCR/Reports/Pages/default.aspx, accessed August, 2014.

29 Later NHGRI created its own cell repository at Coriell and took over direct control of the policies regulating use of the samples.

30 Personal communication from Jean McEwan to ELSI Council Meeting, September 2002.

31 Ibid.

32 For others at NHGRI, the term was meant to differentiate the approach from the formal consultation process required when working with federally recognized tribes. Correspondence with author, November 1, 2016.

33 Interview with author, November 10, 2005.

34 Self-identification based on personal experience emerged within the United States as an authoritative mode of defining oneself in the 1960s. Part of what gave this method its political import was that it relocated the power to define human beings, moving it from governmental authorities and scientific experts back to the people. It responded to the problems of power that intellectuals and activists of the 1960s and 1970s identified with so-called expert production of knowledge about marginalized subjects, such as "women" and "nonwhites." The hope was that this mode of formulating identity would ensure that the subjective dimensions of human beings would not be lost, and thus human beings would not be reduced and treated as mere objects, as in the Tuskegee case, valuable only for the knowledge that could be extracted from them. It is easy to understand why self-identification came to be understood as a democratic victory: it represented a simple example of people overcoming authorities and governing themselves, in this, the most intimate of domains—identity.

35 Interview with author, November 10, 2005.

36 Macer, "Ethical Considerations."

37 Skrentny, *Minority Rights Revolution.*

38 Interview with author, October 27, 2006.

39 Interview with author, June 12, 2006.

40 Interview with author, November 21, 2005.

41 See "NCBI retiring HapMap Resource," http://hapmap.ncbi.nlm.nih.gov/hapmappopulations.html, accessed August 20, 2014.

42 Interview with author, May 30, 2005.
43 Swedlund, "Is There an Echo in Here? Historical Reflections on the Human Genome Diversity Project."
44 International HapMap Consortium, "The International HapMap Project," 793. The adverb *largely* allows these researchers to hold on to a "scientific" concept of race. For more on the strategies genome scientists use to deny the validity of race, while still deploying it, see Reardon, *Race to the Finish: Identity and Governance in an Age of Genomics*.
45 See "Guidelines for referring to the HapMap populations in publications and presentations," http://hapmap.ncbi.nlm.nih.gov/citinghapmap.html.en, accessed December 15, 2015.
46 Interview with author, September 5, 2006.
47 Interview with author, May 30, 2005.
48 International HapMap Consortium, "Integrating Ethics and Science in the International HapMap Project," 467–75.
49 Interview with author, May 30, 2005.
50 International HapMap Consortium, "Integrating Ethics and Science in the International HapMap Project," 470.
51 Interview with author, November 20, 2006.
52 Ibid.
53 Ibid.
54 Butler, *Giving an Account of Oneself*, 22.
55 Interview with author, September 10, 2013.
56 For a full account of this meeting, see Reardon, *Race to the Finish: Identity and Governance in an Age of Genomics*, 74.
57 International HapMap Consortium, "Integrating Ethics and Science in the International HapMap Project," 789–93.
58 Interview with author, November 20, 2006.
59 Interview with author, September 13, 2006.
60 For a discussion of a *thing* as a matter of concern, see chapter 1.
61 Interview with author, November 20, 2006.
62 Interview with author, June 12, 2006.
63 Ibid.
64 Epstein, *Inclusion: The Politics of Difference in Medical Research*.
65 Interview with author, November 20, 2006.
66 Interview with author, June 12, 2006.
67 Reddy, "Good Gifts for the Common Good: Blood and Bioethics in the Market of Genetic Research," 429–72, 438.
68 Ibid., 454.
69 Interview with author, June 12, 2006.
70 Recall that HapMap leaders decided after the initiative began to not collect samples in the United States for the first official phase of the HapMap. For

example, instead of collecting the "Southern European" samples in an Italian American community in New Rochelle, New York, they decided to go to Italy to do the collection. Blood samples were collected in Houston, Texas, and Denver, Colorado, but were not included in the results of the officially celebrated International HapMap Project.

71 Interview with author, November 21, 2005.

72 Ibid.

73 Petryna, *Life Exposed: Biological Citizens after Chernobyl*; Rose and Novas, "Biological Citizenship."

74 Interview with author, September 13, 2006.

75 Council for International Organizations of Medical Sciences (CIOMS), International Ethical Guidelines for Biomedical Research Involving Human Subjects, in collaboration with WHO, http://www.cioms.ch/frame_guidelines_nov_2002.htm, accessed August 4, 2014.

76 Interview with author, May 30, 2005.

77 Interview with author, August 11, 2006.

78 Interview with author, September 11, 2006.

79 Kitcher, *Science, Truth, and Democracy*; Sclove, *Democracy and Technology*.

80 Hacking, *Social Construction of What?*

81 Interview with author, September 11, 2006.

CHAPTER FIVE

1 Fears and Poste, "Building Population Genetics Resources Using the U.K. NHS," 267–68; Fletcher, "Field of Genes: The Politics of Science and Identity in the Estonian Genome Project," 3–14; Gordon, "Why Estonia? A Population's Self-Selection for National Genetic Research"; Marshall, "Tapping Iceland's DNA," 566.

2 Smith et al., "Generation Scotland: The Scottish Family Health Study; a New Resource for Researching Genes and Heritability," 74.

3 See http://www.generationscotland.org/index.php?option=com_content&view=article&id=33&Itemid=35, accessed November 9, 2013. For the Generation Scotland outline, see http://www.generationscotland.org/index.php?option=com_content&view=article&id=33&Itemid=35, accessed November 9, 2013.

4 The notion of public sovereignty dates back to the Enlightenment period when political philosophers, most notably Jean-Jacques Rousseau, argued that the ability to make laws rests with the people. The validity of these laws arose from their expression of the general will of the people. Rousseau, *Discourses on Political Economy and the Social Contract*.

5 Benjamin, "A Lab of Their Own: Genomic Sovereignty as Postcolonial Science Policy," 341–55.

6 Hazell, *The State and the Nations: The First Year of Devolution in the United Kingdom*.

7 Anyone who has ever tried to use Scottish pounds to pay for a taxi in England will know that this is a live and sometimes contentious issue. Consider also the debates about whether citizens of England should be able to go to Scottish universities for free. Jack, "Free Education for Young Scots? Only If the English Students Pay Full Whack."

8 Jasanoff, "Idiom of Co-Production"; Reardon, "Human Genome Diversity Project: A Case Study in Coproduction," 357–88.

9 Specifically, the Genomics Forum was funded by the Economic and Social Science Research Council (ESRC) to ensure social science and public perspectives informed the formation of novel genomic sciences and technologies in the United Kingdom. See "Genomics Forum: ESRC Genomics Policy and Research Forum," GenomicsNetwork, accessed December 20, 2015, http://www.genomicsnetwork.ac.uk/forum/.

10 In 2008, I conducted semistructured interviews with nine of the central researchers and government officials involved in the planning and administration of GS. I also toured the laboratory where the bloods for GS were processed. I reinterviewed many of these same individuals in 2010 and 2012 and retoured the lab. In 2013, I took part in a workshop partially to discuss the future of GS, and in 2014 I returned and gave a seminar based on this chapter that many involved in GS attended, and provided very helpful feedback.

11 Busby and Martin, "Biobanks, National Identity and Imagined Communities: The Case of the UK Biobank," 237–51.

12 For the tensions created in scientific communities, as well as ongoing concerns in the post–World War II era, see Kevles, *In the Name of Eugenics: Genetics and the Uses of Human Heredity*; Stepan, *The Hour of Eugenics: Race, Gender, and Nation in Latin America*; Muller-Hill, *Murderous Science: Elimination by Scientific Selection of Jews, Gypsies, and Others, Germany 1933–1945*. For how the eugenics debate played out in the United Kingdom, and the place of Scotland in these discussions, see Ramsden, "A Differential Paradox: The Controversy Surrounding the Scottish Mental Surveys of Intelligence and Family Size," 109–34.

13 Fletcher, "Field of Genes: The Politics of Science and Identity in the Estonian Genome Project," 3–14.

14 Guerrero Mothelet and Herrera, "Mexico Launches Bold Genome Project."

15 Specter, "Decoding Iceland," 40–51; Rose and Rose, Genes, Cells, and Brains: The Promethean Promise of the New Biology.

16 For an excellent in-depth analysis of the case of Iceland DeCODE Genetics, see Fortun, *Promising Genomics: Iceland and DeCODE Genetics in a World of Speculation*.

17 Winickoff, "Genome and Nation: Iceland's Health Sector Database and Its Legacy," 80–105.

18 At the time, this was the largest deal ever struck between a major pharmaceutical company and a genomics company. See http://www.thefreelibrary

.com/Hoffmann-La+Roche+and+deCODE+genetics+Sign+Genomics+Collaboration+To . . .-a020209570, accessed April 10, 2013. La Roche did stipulate that certain benchmarks needed to be met for it to pay the full amount of funds. deCODE failed to meet these benchmarks, and in the end only received about $74.3 million. For an analysis, see Sigurdsson, "Springtime in Iceland and 200 Million Dollar Promises."

19 Icelandic biologists and medical scientists questioned this assumption of homogeneity. See Zoëga and Andersen, "Icelandic Health Sector Database: Decode and the 'New' Ethics for Genetic Research."

20 For a discussion of the struggles over the meaning of what it means to be educated and informed about human genetic variation research, see Reardon, *Race to the Finish: Identity and Governance in an Age of Genomics.*

21 Interview with author, February 12, 1999.

22 Specter, "Decoding Iceland," 49.

23 Duncan, "World Medical Association Opposes Icelandic Gene Database," 1096; Eyfjörd, Ögmundsdottir, and Zoëga, "Decode Deferred," 491.

24 Specter, "Decoding Iceland," 44.

25 Lewontin, "People Are Not Commodities."

26 Specter, "Decoding Iceland," 49–50.

27 Appiah, *Cosmopolitanism: Ethics in a World of Strangers.*

28 See http://www.nytimes.com/2000/06/27/science/reading-the-book-of-life-white-house-remarks-on-decoding-of-genome.html?pagewanted=all&src=pm, accessed June 25, 2013; Wade, "Scientists Complete Rough Draft of Human Genome"; Sulston and Ferry, *Common Thread: A Story of Science, Politics, Ethics, and The Human Genome.*

29 As an early leader of Generation Scotland explained to me: "My concept was Scotland: single healthcare provider, everybody has welfare health provision, very little private medical care, essentially a closed population. Although we are part of mainland UK, where for historical and cultural reasons we operate as a nation within a nation, people in Scotland tend to stay put. It's therefore possible to track the health records of individuals from grandparents to parents to children and their children. And by virtue of the relative stability of that population, it's remarkably homogeneous, with 98% of northern European Caucasian origin. . . . Ninety-eight percent." Interview with author, November 26, 2008.

30 McKie, "Steam Rises over Iceland Gene Pool"; Meek, "Decode Was Meant to Save Lives . . . Now It's Destroying Them."

31 Hickman, "GM Crops: Protesters Go Back to the Battlefields."

32 See Boseley and Carter, "He Stripped the Organs from Every Dead Child He Touched." The scandal eventually led to the passage in the United Kingdom of the Human Tissue Act of 2004 and the creation of the Human Tissues Authority.

33 In 1999, the Royal Society created the Science in Society program that sought to innovate approaches to engaging the public in the design and governance of research. In 2000, the House of Lords Science and Technology Select Committee recommended that "direct dialogue with the public . . . should become an integral part of the [science] process." House of Lords, "Science and Society." Generation Scotland quotes this section of the House of Lords report on its website. See http://www.generationscotland.org/index.php?option=com_content&view=article&id=61&Itemid=126, accessed April 10, 2013.

34 Indeed, I first learned of Generation Scotland because I had been invited to give a plenary talk at the launching of the UK Economic and Social Science Research Council's Genomics Policy and Research Forum in Edinburgh, Scotland, a new institution designed to connect social science research on the social and economic implications of genetics discoveries with public and policy debate. See also the Royal Society, *Science in Society: The Impact of the Five Year Kohn Foundation Funded Programme*.

35 See http://www.generationscotland.org/index.php?option=com_content&view=article&id=61&Itemid=126, accessed April 10, 2013.

36 Wakeford and Hale, *Generation Scotland: Towards Participatory Models of Consultation*.

37 Interview with author, July 10, 2008.

38 See http://www.generationscotland.org/index.php?option=com_content&view=article&id=61&Itemid=126, accessed April 10, 2013.

39 One of the GS publications, *Generation Scotland: Towards Participatory Models of Consultation*, invokes the ideas of Jean-Jacques Rousseau, the father of social contract theory, to justify the participation of citizens in decisions about GS. For Rousseau's views on the social contract, see Rousseau, *Discourses on Political Economy and the Social Contract*, 18.

40 From this screen, a "Discover Generation Scotland" button appears, a button that brings up an overview of GS with a header that reads "Addressing the Health and Wealth of Scotland." Below the header one learns that "Generation Scotland will be to the medical, social and economic benefit of Scotland and its people." http://129.215.140.49/gs/gindex.html, accessed December 16, 2008.

41 See "Health in Scotland," Scottish Government, accessed December 14, 2008, http://www.scotland.gov.uk/Publications/2004/04/19128/34905. Further, in 2003, a report issued by David Leon and colleagues at the London School of Hygiene and Tropical Medicine cited death rates among Scots of working age (fifteen to seventy-four years) as the highest in western Europe. See http://www.scotpho.org.uk/home/Comparativehealth/InternationalComparisons/int_mortality_comparisons.asp, accessed December 14, 2008.

42 Generation Scotland's focus on mental health, a national public health priority in Scotland, particularly positioned GS as a project of and for the people of Scotland. For an explanation of the current public health focus on mental health in Scotland, see "Mental Health Indicators," NHS Health Scotland, accessed December 14, 2008, http://www.healthscotland.com/scotlands-health/population/mental-health-indicators.aspx.

43 Interview with author, July 17, 2012.

44 See "Scots Family Study Logs 1000th Volunteer," University of Dundee, accessed April 10, 2013, http://www.dundee.ac.uk/pressreleases/2006/prdec06/familystudy.html.

45 There is a prevalent notion among analysts of Scottish government that as a small nation, getting things done requires working collectively and in partnerships with the nation's people. Keating, *Government of Scotland: Public Policy Making after Devolution*, 216. It is common practice to conduct "public consultations" to seek input from citizens on national policy and planning issues. Generation Scotland would be no different. Haddow et al., *Public Attitudes to Participation in Generation Scotland.*

46 Davie, *The Democratic Intellect: Scotland and Her Universities in the Nineteenth Century.*

47 See http://129.215.140.49/gs/ukb.htm, accessed December 14, 2008. Scots do indeed have a long history of participating in genetic studies going back eighty years to the 1932 cohort study. Interview with author, December 2, 2008.

48 See http://www.generationscotland.org/images/stories/RSE_meetingreport_-_aressing_the_health_and_wealth_of_scotland.pdf, accessed April 25, 2013.

49 Chief Scientist Office, Guidance for Applicants to the Genetics and Health Care Initiative, 2.

50 See http://www.tmri.co.uk/news/item.aspx?a=25&z=1, accessed April 15, 2013.

51 Anderson et al., "Effects of Estrogen Plus Progestin on Gynecologic Cancers and Associated Diagnostic Procedures: The Women's Health Initiative Randomized Trial," 1739–48.

52 Brower, "A Second Chance for Hormone Replacement Therapy?," 1112–15.

53 Interview with author, December 2, 2008.

54 Dawson, *Public Perceptions on the Collection of Human Biological Samples.*

55 See, for example, book I, chapter 8, of the *Wealth of Nations* in which Smith writes: "But what improves the circumstances of the greater part can never be regarded as an inconveniency to the whole. No society can surely be flourishing and happy, of which the far greater part of the members are poor and miserable. It is but equity, besides, that they who feed, cloath and lodge the

whole body of the people, should have such a share of the produce of their own labour as to be themselves tolerably well fed, cloathed and lodged." See conlib.org/library/Smith/smWN3.html#I.8.35, accessed April 10, 2013.

56 In a well-known passage from the *Wealth of Nations* titled "Of the Origin and Use of Money," Smith describes the practical need for a common commodity, or currency: "One man, we shall suppose, has more of a certain commodity than he himself has occasion for, while another has less. The former consequently would be glad to dispose of, and the latter to purchase, a part of this superfluity. But if this latter should chance to have nothing that the former stands in need of, no exchange can be made between them. . . . In order to avoid the inconveniency of such situations, every prudent man in every period of society, after the first establishment of the division of labour, must naturally have endeavoured to manage his affairs in such a manner as to have at all times by him, besides the peculiar produce of his own industry, a certain quantity of some one commodity or other, such as he imagined few people would be likely to refuse in exchange for the produce of their industry." See "An Inquiry into the Nature and Causes of the Wealth of Nations," Adam Smith, accessed April 10, 2013, http://geolib.com/smith.adam/won1-04.html.

57 Rousseau, Discourses on Political Economy and the Social Contract.

58 Haddow et al., "Generation Scotland: Consulting Publics and Specialists at an Early Stage in a Genetic Database's Development," 139–49.

59 Interview with author, December 2, 2008.

60 This would later change as GS took on a more streamlined governance model and created an executive committee that contained no social scientists.

61 Haddow et. al., "Generation Scotland: Consulting Publics and Specialists at an Early Stage in a Genetic Database's Development," 139–49.

62 Social scientists involved in GS critiqued other public consultation efforts for not following up their findings with changes to policy. Without this follow-up, they argued, these exercises proved mere window dressing that led to little substantive change. Ibid.

63 As one of the natural and medical scientists leading GS explained: "It's important that those ideas that are developed through a dialogue with the social scientists are tested using appropriate instruments, social science measures and instruments." Interview with author, November 27, 2008.

64 Ibid.

65 Haddow et al., Public Attitudes to Participation in Generation Scotland, 276.

66 Ibid.

67 Importantly, as we will see later, participants also preferred that pharmaceutical companies not be given access to their DNA samples. Ibid., 31.

68 Ibid.

69 See http://www.genomicsnetwork.ac.uk/esrcgenomicsnetwork/news/media room-pressreleases/2005pressreleases/title,91,en.html, accessed April 10, 2013. As one social science researcher working on the project reflected, this commitment was real. When faced with the hard questions, GS leadership committed to responding to public input. Interview with author, July 18, 2012.

70 Generation Scotland, Generation Scotland Management, Access, and Policy Publication.

71 Interview with author, November 26, 2008.

72 Interview with author, July 17, 2012.

73 Haddow et al., Public Attitudes to Participation in Generation Scotland, 279.

74 Interview with author, July 18, 2012.

75 Interview with author, November 27, 2008. For a discussion of how biobanks in countries around the globe have differently imagined and enacted notions of property, benefits, and democratic governance of biological tissues, see Gottweis and Peterson, *Biobanks: Governance in Comparative Perspective*; Kaye and Stranger, *Principles and Practice in Biobank Governance.*

76 Haddow et al., "Public Attitudes to Participation in Generation Scotland," 280.

77 Interview with author July 17, 2012.

78 Interview with author, July 18, 2012.

79 Interview with author, November 26, 2008.

80 Interview with author, March 8, 2012.

81 Interview with author, July 10, 2008.

82 Laurie and Gibson, *Generation Scotland: Legal and Ethical Aspects.* As a way to respond to potential resentment arising from research subjects not receiving profits or direct benefits while being subjected to possible breaches of privacy, the authors of the report recommend "reliance on Scottish identity and the 'Scottishness' of GS to engender feelings of ownership of the project." Laurie and Gibson, *Generation Scotland: Legal and Ethical Aspects*, 39.

83 As one GS organizer explained: "There's a strong sense of national pride and, and there was a thought that if we said 'Hey, we will take these samples to England,' for example, that people might be less likely to say yes." Interview with author, July 17, 2012.

84 Ibid.

85 There is also historical precedence for this approach. As Helen Busby and Paul Martin have argued, mobilizing people on the basis of national citizenship to take part in blood drives carried with it the expectation that the blood would be used in the United Kingdom. Busby and Martin, "Biobanks,

National Identity, and Imagined Communities: The Case of UK Biobank," 242.

86 Reich and Lander, "On the Allelic Spectrum of Human Disease," 124–37.

87 Dickson et al., "Rare Variants Create Synthetic Genome-Wide Associations," 1.

88 Angrist, *Here Is a Human Being*.

89 Interview with author, July 18, 2012.

90 Interview with author, March 8, 2012.

91 Interview with author, March 8, 2012.

92 Interview with author, July 18, 2012.

93 As I was told in 2008: "I have to be honest, in this particular second phase, there are no exemplars at all [of public engagement and education]. It is just trying to get as many people in [to participate in GS] as possible." Interview with author, December 2, 2008.

94 Interview with author, July 18, 2012.

95 Ibid., 20.

96 Interview with author, July 17, 2012.

97 Interview with author, July 18, 2012.

98 As a member of the GS team explained to me: "They are increasingly finding that when they do more and more detailed analysis . . . the parents don't have the mutation . . . so that mutation in the child has occurred de novo. It's not a hundred percent of situations, far from it, but it's like maybe 40 percent. And in a way when you think about it, it's not in the least bit surprising because lots of things go wrong in human reproduction, and most of the time, probably they are lethal. I think I read some paper years ago about the number of spontaneous abortions where a woman barely knows she's pregnant. It gets to about two weeks, and then ping. I think we are pretty inefficient at having babies." Interview with author, July 18, 2012.

99 Interview with author, July 17, 2012.

100 Ibid.

101 Interview with author, March 8, 2012.

102 See "SFHS Participants Continue to Support Generation Scotland," Generation Scotland, accessed April 4, 2014, http://www.generationscotland.co.uk/index.php?option=com_content&view=article&id=86:sfhs-participants-continue-to-support-generation-scotland&catid=39:latest-news&Itemid=5.

103 Benjamin, "A Lab of Their Own: Genomic Sovereignty as Postcolonial Science Policy," 341–55.

104 Appiah, *Cosmopolitanism: Ethics in a World of Strangers*; Butler, *Precarious Life: The Powers of Mourning and Violence*.

105 See, for example, the recent Nuffield Council on Bioethics report, "Solidarity: A Reflection on an Emerging Concept in Bioethics," accessed November 7, 2013, http://www.nuffieldbioethics.org/solidarity-0. See also Prainsack and Buyx, *Solidarity and Biomedicine and Beyond*.

106 Interview with author, November 26, 2008.

107 Donating blood and other human tissues for altruistic reasons has a long and much-celebrated history in the United Kingdom. See Titmuss, *Gift Relationship: From Human Blood to Social Policy*. However, as we will see both in this and subsequent chapters, invoking altruism in the context of genomics raised questions about the changing ethical, economic, and political contexts of tissue exchange.

108 Arendt, *Between Past and Future*, 185. I will return to and expand on this point in chapter 7.

109 Personal communication with Robin Morton and Archie Campbell, February 13, 2017.

110 Hartocollis, "Cancer Centers Racing to Map Patients' Genes."

CHAPTER SIX

1 For a critical appraisal of this historical moment, and the ideology of early social media, see Van Dijck, *Culture of Connectivity: A Critical History of Social Media*, 11.

2 Notable among the others that will be a part of my story are Navigenics, deCODEme, and Pathway Genomics.

3 See http://mediacenter.23andme.com/blog/2008/09/09/23andme-democratizes-personal-genetics/, accessed December 17, 2015.

4 Avey, "Inspiration Comes in Many Forms."

5 Pollack, "The Wide, Wild World of Genetic Testing."

6 The Department of Public Health in New York sent similar letters to twenty-six companies six months prior. Wadman, "Gene-Testing Firms Face Legal Battle," 1148–49.

7 Pollack, "F.D.A. Faults Companies on Unapproved Genetic Tests," B2.

8 They are only of value, the filing further explains, to 23andMe for it is only the company—and not people—that has the power to make something of the test data—namely, "databases and statistical information" that it "markets to other sources." See http://dockets.justia.com/docket/california/casdce/3:2013cv02847/429459, accessed October 12, 2014.

9 Arendt, *The Human Condition*, 5.

10 MacKenzie, "Machine Learning and Genomic Dimensionality," 73–102.

11 See, for example, Jenny Reardon and Jacob Metcalf, "Biocuration Workshop Report," accessed December 22, 2015, http://scijust.ucsc.edu/wp-content/uploads/2013/04/Biocuration-Workshop-Report.sjwgrapp.pdf.

12 Specifically, the chapter is based on sixteen semistructured interviews conducted between 2008 and 2012 with those who founded or worked at PG companies, and academics who worked on the questions of biocuration raised by PG. Between 2008 and 2010, I also visited and toured the company headquarters of two personal genetics companies, and other bioinformatics

companies who worked on biocuration. These visits afforded me the chance to learn more informally from company employees and to get a feel for company cultures.

13 Rabinow, "Artificiality and Enlightenment: From Sociobiology to Biosociality."

14 Thomas Goetz offers this description: "Brin proposes a different approach, one driven by computational muscle and staggeringly large data sets. It's a method that draws on his algorithmic sensibility—Google's storied faith in computing power—with the aim of accelerating the pace and increasing the potential of scientific research. . . . In other words, Brin is proposing to bypass centuries of scientific epistemology in favor of a more Googley kind of science. He wants to collect data first, then hypothesize, and then find the patterns that lead to answers. And he has the money and the algorithms to do it." Goetz, "Sergey Brin's Search for a Parkinson's Cure."

15 Avey writes: "A seminal component of genetics research is, and has always been, the so-called human subject, a real, live person who contributes his or her highly valuable information. . . . By establishing, and sometimes vociferously defending, the firewall between the subjects and themselves—with an intervening IRB-approved consent form typically guaranteeing no personal benefit back to the participants—researchers have been absolved of, or disallowed, as the case may be, any responsibility or ability to communicate directly with their subjects (sounds almost feudal, doesn't it?), as if post-study phenotypic details are irrelevant to the task at hand." Avey, "Inspiration Comes in Many Forms."

16 That is, anyone who lives in a jurisdiction that allows 23andMe to sell its kits. 23andMe hoped to have a global reach, but the legal issues of selling in other countries posed significant barriers. In the United States, in 2008 New York and California sent cease and desist letters to 23andMe. Pollack, "California Licenses 2 Companies to Offer Gene Services."

17 23andMe company culture exudes childlike playfulness. Corporate headquarters and the website are painted in Crayola colors. When I visited, there were toys in the bathroom, and I left with stickers and a T-shirt. I had hoped for a beanie, but they no longer had these.

18 Avey, "Inspiration Comes in Many Forms."

19 Interview with author, September 9, 2011.

20 Ibid.

21 See http://blog.23andme.com/2008/02/27/what-you-can-do-for-23andme-and-future-generations/, accessed October 6, 2014.

22 http://blog.23andme.com/23andme-and-you/anne-and-linda-unveil-23andwe-at-d6/, accessed September 10, 2014; "23andWe Research Update: Something Old and Something New," *23andMe Blog*, December 10, 2010, accessed

September 10, 2014, http://blog.23andme.com/23andme-research/23andwe-research-update-something-old-and-something-new/.

23 Participants received the 23andMe service for free. "The Search for a Cure Starts with Your DNA," 23andMe, accessed October 2, 2014, https://www.23andme.com/pd/.

24 http://mediacenter.23andme.com/press-releases/23andme-and-the-parkinsons-institute-announce-initiative-to-advance-parkinsons-disease-research/, accessed September 10, 2014.

25 Pollack, "Google Co-Founder Backs Vast Parkinson's Study."

26 See https://www.23andme.com/you/survey/take/health_intake/finish/, accessed September 10, 2014.

27 Interview with author, September 9, 2011.

28 "Introducing a Do-It-Yourself Revolution in Disease Research," 23andMe, accessed September 22, 2014, http://blog.23andme.com/23andme-and-you/introducing-a-do-it-yourself-revolution-in-disease-research/.

29 Turner, *From Counterculture to Cyberculture*, 103–7.

30 Sunder Rajan, *Biocapital: The Constitution of Postgenomic Life*, 283.

31 Gertner, *The Idea Factory: Bell Labs and the Great Age of American Innovation.*

32 Interview with author, September 9, 2011.

33 For a critique of this idealistic view of the moral economies of genetics and genomic research, I refer the reader to chapter 2.

34 Interview with author, September 9, 2011.

35 Ibid.

36 Anderson, *Free: The Future of a Radical Price.*

37 Winterhalter, "A Genetic 'Minority Report': How Corporate DNA Testing Could Put Us at Risk."

38 Interview with author, September 9, 2011.

39 Yet even those who are the biggest fans of these powers of convergence, such as media scholar Henry Jenkins, point out that convergence is not all about new freedoms for the people; it is also about new forms of consumption and power for corporations. However, at the end of the day Jenkins suggests that consumption is not as opposed to democratic citizenship as we thought. Maybe we can vote with our dollar. And perhaps participation in commodified popular culture "prepar[es] the way for a more meaningful popular culture." Jenkins, *Convergence Culture: Where Old Media and New Media Collide*, 222. This might have sounded like heresy a decade ago, but increasingly—partly because of companies we love, such as Apple—the ideas are articulable; indeed, they sell.

40 These words begin the Google Code of Ethics. See https://investor.google.com/corporate/code-of-conduct.html, accessed September 12, 2014.

41 For the first news article on the PGP, see Toner, "Harvard's Church Calls for 'Open Source,' Non-Anonymous Personal Genome Project."

42 This was particularly true of NHGRI.

43 The HGP only produced a haploid sequence. For a description of the Venter genome, see Levy et al., "The Diploid Genome Sequence of an Individual Human."

44 Kennedy, "Not Wicked, Perhaps, but Tacky," 1237.

45 The full transcript of this interview can be found at http://www.yourgenome.org/people/james_watson.shtml, accessed September 12, 2014.

46 Interview with author, September 9, 2011.

47 Interview with author, June 15, 2010.

48 When 23andMe did market, it did so in a distinctive way, placing a "decal" on a distinguished piece of technology, the dirigible. According to Airship Enterprises, the host of the 23andMe decal, which took six days to affix, decals represented the latest advance in marketing: "Special terms have become the trademark of technology and business. What used to be 'Advertising' has now become 'Branding' and what used to be 'Billboards' are in Airship Ventures vernacular 'Decals.' The 'Eureka's' first decal was Pixar™'s 'UP' lettering and the wonderful depiction of the main character's house being airlifted by hundreds of colorful balloons to its new home far away. The second decal, currently on the 'Eureka,' is '23and Me.com Personal Genetics.' It invites interested parties and individuals to access the company's website to see their genes in a whole new light." See http://www.nasa.gov/centers/ames/pdf/576856main_summer09.pdf, accessed September 22, 2014.

49 Interview with author, June 15, 2010.

50 Salkin, "When in Doubt, Spit It Out."

51 On distinction and habitus, see Bourdieu, *Distinction: A Social Critique of the Judgement of Taste*.

52 Interview with author, June 15, 2010.

53 While this assumption is now mainstream, there are several critiques of the relevance of race, gender, and class to genomic interpretation. See, for example, Bolnick et al., "The Science and Business of Genetic Ancestry Testing," 399–400.

54 Interview with author, June 15, 2010.

55 Lowe, "Google Wife Targets World DNA Domination."

56 23andMe, "Time to Thank Our Friends."

57 Avey, "Inspiration Comes in Many Forms."

58 See "What Are Genes," 23andMe, accessed September 15, 2014, https://www.23andme.com/gen101/genes/.

59 23andMe, *23andMe DNA Processing Lab Video* (2014), http://www.youtube.com/watch?list=PLFvkfbvHpp1lEK7U0ehu_akvIX9ZIzJDc&v=0gC8RQ7PemM#t=16.

60 Wojcicki, "Research Participants Have a Right to Their Own Genetic Data."

61 Similarly, Misha Angrist, member of the ELSI Advisory Board of the nonprofit Personal Genome Project, describes personal genetics as part of the "citizen science" movement. Angrist, "Eyes Wide Open: The Personal Genome Project, Citizen Science, and Veracity in Informed Consent," 691–99.

62 Interview with author, September 9, 2011.

63 Genome Web, "Walgreens to sell Pathway Genomics' Sample Collection Kit." By 2013, 23andMe also was heavily marketing to a broad audience, with commercials on stations such as the Food Channel. See http://www.ispot.tv/ad/7qoF/23-and-me, accessed October 3, 2014.

64 Shuren, "Statement before the Subcommittee on Oversight and Investigations, Committee on Energy and Commerce, United States House of Representatives."

65 Ibid. While congressmen raised these fears, geneticists conducted research attempting to demonstrate that such fears were overblown. See Green et al., "Disclosure of *APOE* Genotype for Risk of Alzheimer's Disease," 245–54.

66 Locke, *Two Treatises of Government.*

67 Rousseau, *Discourses on Political Economy and the Social Contract.*

68 I heard similar fears articulated in the United Kingdom, where strains on the already stretched National Health Service was a particular concern. Interview with author, July 16, 2008.

69 Shurin, "Statement before the Subcommittee on Oversight and Investigations, Committee on Energy and Commerce, United States House of Representatives."

70 Interview with author, August 9, 2010.

71 In 2008 and 2009, 23andMe reported receiving numerous requests from journalists to take part in their service. While this provided them with much free publicity, the report of different results created a credibility problem. See, for example, Davies, "Keeping Score of Your Sequence,; Hunter, "Letting the Genome out of the Bottle—Will We Get Our Wish?," 105–7.

72 Ng, "An Agenda for Personalized Medicine," 724–26.

73 As the historian of science Lorraine Daston has documented, since at least the seventeenth century, objectivity "has been a monolithic and immutable concept" that assumes that only by restraint of emotions and personal views can objective truth be discerned. In short, objectivity is impersonal and aperspectival. See Daston, "Objectivity and the Escape from Perspective," 110–23.

74 Personalized Medicine Coalition, "Personal Genomics and Industry Standards: Scientific Validity," 1–4.

75 Interview with author, June 15, 2010.

76 Interview with author, March 30, 2010.

77 They hoped this information would shape customers' interactions with the health care system. For example, someone who receives a Navigenics report indicating a higher risk of colon cancer might choose not to wait until they get older to have a colonoscopy. This approach proved attractive to employee wellness programs that bought into Navigenics and who hoped that employees who received genomic information would make healthier choices.

78 See the comments section of 23andMe, "Google Co-Founder Blogs about 23andMe Data, Parkinson's Risk."

79 Interview with author, March 30, 2010.

80 Haraway, *Simians, Cyborgs, and Women*, 193.

81 Interview with author, October 24, 2010.

82 Williams, "On the Tip of Creative Tongues: Curate."

83 Strasser, "Experimenter's Museum: Genbank, Natural History, and the Moral Economies of Biomedicine, 1979–1982," 60–96.

84 Interview with author, October 24, 2010.

85 Darnovsky, "Spitterati and Genome Research; Winterhalter, "A Genetic 'Minority Report': How Corporate DNA Testing Could Put Us at Risk."

86 Author's field notes, October 5, 2009.

87 Ibid.

88 House of Representatives, Subcommittee on Oversight and Investigations, Committee on Energy and Commerce. Direct to Consumer Genetic Testing and the Consequences to the Public Health; Darnovsky, "The Spitterati and Genome Research."

89 Interview with author, October 24, 2010.

90 For an in-depth study of the ways in which the boundaries of private and academic science blur, and a history of these relations, see Shapin, *Scientific Life: A Moral History of a Late Modern Vocation.*

91 For a copy of the letter that the FDA sent to 23andMe, see Alberto Gutierrez, e-mail message to author, October 6, 2014, http://www.fda.gov/iceci/enforcementactions/warningletters/2013/ucm376296.htm.

92 See http://blog.23andme.com/news/23andme-provides-an-update-regarding-fdas-review/, accessed October 6, 2014.

93 See Wojcicki, "23andMe Provides an Update Regarding FDA's Review."

94 For example, Navigenics, bought by Life Technologies in 2012, informed its customers that it planned to destroy all customer genetic results and samples in August of 2015. See http://thegenesherpa.blogspot.com/2012/08/5-years-later-navigenics-fulfills-my.html, accessed November 16, 2016.

95 Interview with author, November 18, 2010.

96 See http://blog.23andme.com/2008/02/27/what-you-can-do-for-23andme-and-future-generations/, accessed October 6, 2014.

97 Interview with author, March 19, 2010.

98 The conservative nature of NIH-funded approaches is a frequent theme of those who promote personal genomics. One scientist I talked with who works in the field told me that some argue that NIH stands for "Not Invented Here." Author field notes.

99 Interview with author, September 3, 2010.

100 Ibid.

101 In its November 2013 letter to 23andMe, the FDA included a lengthy paragraph explaining how many e-mails, face-to-face meetings, and time and energy it had put into working with 23andMe. The implication was that 23andMe failed to reciprocate with its time and energy. See Alberto Gutierrez, e-mail message to author.

102 Lowe, "Google Wife Targets World DNA Domination." For a similar description of Google, see Kelleher, "Google: World Domination Starts Today."

103 Cook-Deegan, *Gene Wars: Science, Politics, and the Human Genome*.

104 See, for example, this talk by Wojcicki and Avey, *23andMe Research Revolution* (2014), http://www.youtube.com/watch?v=WfI62N8pOkE.

105 Lyotard, *Postmodern Condition: A Report on Knowledge*, 5.

106 Wrote the Personalized Medicine Coalition: "Transparency in disclosing companies methods and criteria is the most pragmatic goal," Personalized Medicine Coalition, "Personal Genomics and Industry Standards: Scientific Validity," 3.

CHAPTER SEVEN

1 Desmond-Hellman, "Toward Precision Medicine: A New Social Contract?," 129.

2 For this formulation of patients helping doctors, see Desmond-Hellman, "Attention Stressed Out Doc: Can the Consumer Be the 'Cavalry' That Saves You?"

3 Philippidis, "Top 10 Biopharma Clusters."

4 Kelty, *Two Bits: The Cultural Significance of Free Software*, 10, 23, 310.

5 On the life and legacy of Aaron Swartz, see Peters, *The Idealist: Aaron Swartz and the Rise of Free Culture of the Internet*. On the copyright wars, see Baldwin, *Copyright Wars: Three Centuries of Trans-Atlantic Battle*; and Johns, *Piracy: The Intellectual Property Wars from Gutenberg to Gates*.

6 See Reardon, "Should Patients Understand That They Are Research Subjects?" Inside the magazine, the *San Francisco Chronicle* also published a response from UCSF and a supporting editorial.

7 Harris, "Letter to the Editor."

8 As it turned out, my doctor also had concerns. We cannot, she explained to me, get people to take care of themselves and do things even when health

care recommendations are simple, recommended to all, and reinforced through frequent public health messages. What will happen when the system becomes more complex, and medical advice changes constantly based on changing research results?

9 For one exploration of the variety of actors and historical contingencies that led to the tying of research funding to the cause of "fighting disease," see Creager, "Mobilizing Biomedicine: Virus Research between Lay Health Organizations and the U.S. Federal Government, 1935–1955," 171–201.

10 For an excellent discussion of the conditions that gave rise to the life sciences as a space of the in-between, where the exchange of materials constitutes neither "taking" nor "giving," but rather both, see Hayden, "Taking as Giving: Bioscience, Exchange, and the Politics of Benefit-Sharing," 729–58.

11 In addition to coffee breaks and ice cream outings with colleagues, this chapter is based on one round of semistructured interviews with the main organizers of the PGP conducted in 2010, with follow-ups in 2013. I also attended two of the main annual meetings of the PGP, the Genes, Environment, and Traits (GET) conferences, in 2012 and 2014. With their permission, and in the spirit of the openness of the PGP, in this chapter, I use some interviewee names.

12 Angrist, *Here Is a Human Being: At the Dawn of Personal Genomics*; Church and Regis, *Regenesis: How Synthetic Biology Will Reinvent Nature and Ourselves.*

13 Church, "Genomes for All," 47–54.

14 "Life Technologies takes benchtop genome sequencer on tour of UK in a Mini." MTB Europe, http://www.mtbeurope.info/news/2012/1209012.htm, accessed November 16, 2016.

15 During the HGP era, all eyes remained on the prize—the human genome sequence. Almost all money for sequencing—private and public—went toward sequencing the human genome, and only two main efforts existed: the public Human Genome Project and the private effort to sequence the human genome led by Celera. Both used the ABI machines. To the dismay of Church, ABI secured a decade-long monopoly on the sequencing machine industry. Church and Regis, *Regenesis: How Synthetic Biology Will Reinvent Nature and Ourselves*, 167.

16 Ibid.

17 Ibid., 174.

18 Interview with author, October 17, 2010.

19 Lunshof et al., "From Genetic Privacy to Open Consent," 410.

20 Interview with author, October 17, 2010.

21 Ibid.

22 Ibid. Church sought genomes for whole genome sequencing. Whole genome sequence uniquely identifies individuals, and so maintaining anonymity poses significant technical and legal challenges.

23 Church and Regis, *Regenesis: How Synthetic Biology Will Reinvent Nature and Ourselves*, 212.
24 On taking as a form of giving, see Hayden, "Taking as Giving: Bioscience, Exchange, and the Politics of Benefit-Sharing."
25 Church, "Genomes for All," 48.
26 Interview with author, July 24, 2013.
27 For a description of the DIYBio movement that Bobe helped to found, see "An Institution for the Do-It-Yourself Biologist."
28 For leaders of the Personal Genome Project, personal genomics referred to this Do-It-Yourself spirit that came out of the personal computing worlds. As personal genomics became more mainstream and commercial, this reference proved hard to maintain, leading to desires to change the name of the initiative. Today Bobe leads Open Humans, which is essentially an extension of the PGP but with a name and institutional structure that better fits the goals of the initiative.
29 Interview with author, June 4, 2102.
30 I use "boys" here intentionally. I know of no genome sequencing innovator who is not a man.
31 Interview with author, July 24, 2013.
32 As Church explains on his personal website: "From the very beginning of the HGP I was lobbying for improved sequencing technology and against excessive costs of focusing on mapping, on racing with companies, and on cataloging common SNPs." Church, "Human Genome Project (HGP) History (a Personal Account)."
33 Interview with author, October 17, 2010. For more information about GenSpace, see http://genspace.org/, accessed November 16, 2016.
34 Regaldo, "Geneticists Begin Tests of an Internet for DNA."
35 Interview with author, May 24, 2012.
36 Interview with author, July 16, 2012.
37 Interview with author, July 24, 2013.
38 See Timmerman, "Stephen Friend, Leaving High Powered Merck Gig, Lights Fire for Open Source Biology Movement."
39 "Your Data Are Not a Product," 357–57.
40 Interview with author, June 18, 2012.
41 See "Open Humans Network."
42 Reportedly, he also did not think the PGP presented a clear scientific hypothesis.
43 Angrist, *Here Is a Human Being: At the Dawn of Personal Genomics*. The PGP sought funding from NIH as early as 2003, years before the passage of the Genetic Information Nondiscrimination Act (GINA) of 2008. Even after the passage of the law, denial of long-term care and life insurance remained a serious threat.

44 See "Consent Form," Harvard Medical School, accessed January 5, 2015, http://www.personalgenomes.org/static/docs/harvard/PGP_Consent_Approved_02212013.pdf.

45 The PGP leadership argued that veracity and openness of research design should be the leading principles in obtaining participant consent. Lunshof et al., "From Genetic Privacy to Open Consent," 406–11.

46 Indeed, PGP read time data for the forms indicated that this happened frequently. Interview with author, July 25, 2013.

47 See "Consent Form," Harvard Medical School, accessed January 5, 2015, http://www.personalgenomes.org/static/docs/harvard/PGP_Consent_Approved_02212013.pdf. The exam also fostered the democratic ideal of inclusion. To begin with, the Harvard IRB required all PGP participants to have a master's degree in genetics. The entrance exam replaced this requirement, opening the PGP up to the imagined average American. Bobe explained: "The trajectory of the PGP is to go from Journal Club to Sam's Club, right? For the PGP Ten, the IRB requested a Master's degree [in genetics] and that wasn't very fair. Like with me, I don't have a Master's degree in genetics. . . . And so that was the genesis of the entrance exam: let's make it fair." Interview with author, October 17, 2010.

48 Charis Thompson nicely conveys this spirit of the PGP open consent in Thompson, *Good Science: The Ethical Choreography of Stem Cell Research*, 182.

49 Tanner, "Harvard Professor Re-Identifies Anonymous Volunteers in DNA Study." However, it should be noted that Sweeney and Erlich "re-identified" participants in study populations for which much information about the participants was easy to obtain: PGPers who already agreed to place information in the public domain and members of the CEPH population in Utah, many of whom are Mormon and heavily involved in ancestry research. As Malia Fullerton at the University of Washington explained to me, given these particularities of the study population, it is not clear how far the findings about the robustness of privacy extend. See also Gitschier, "Inferential Genotyping of Y Chromosomes in Latter-Day Saints Founders and Comparison to Utah Samples in the HapMap Project," 251–58.

50 Ibid.

51 Interview with author, October 17, 2010.

52 Perlroth, "Ashley Madison Chief Steps Down after Data Breach."

53 Interview with author, July 25, 2013.

54 Ibid.

55 For an excellent account of the passion and resources invested in developing sequencing machine technologies in the opening decade of the twenty-first century, see Angrist, *Here Is a Human Being: At the Dawn of Personal Genomics*.

56 Anderson, *Free: The Future of a Radical Price*.

57 ABI merged with Invitrogen in 2008 to create Life Technologies. Applera no longer exists.
58 Interview with author, July 24, 2013.
59 See "Personal Genome Project (PGP)," Coriell Institute for Medical Research, accessed August 25, 2015, https://catalog.coriell.org/0/Sections/Collections/NIGMS/PGPs.aspx?PgId=772&coll=GM.
60 This lack of diversity is linked to—among other problems—inaccurate genetic test results. See, for example, Manrai et al., "Genetic Misdiagnoses and the Potential for Health Disparities," 655–65.
61 Interview with author, July 24, 2013.
62 Ibid.
63 Bobe opened his talk with an image from the American Continental Congress. Field notes, June 14, 2011. For a description of the event, see "Codes," DIYBio, accessed August 25, 2015, http://diybio.org/codes/.
64 As Benjamin Franklin wrote: "Only a virtuous people are capable of freedom." Franklin quoted in Foner, *Story of American Freedom*, 8.
65 Author field notes, April 25, 2014. New York Times Editorial Board, "Silicon Valley's Diversity Problem," SR10.
66 As the American historian Eric Foner argues, the obvious absence of so many from the rights of citizenship drove freedom movements in the United States. See Foner, *Story of American Freedom*.
67 Author field notes, April 25, 2012.
68 Ibid.
69 The informed consent form for the PGP announces: "The Personal Genome Project is a new form of public genomics." See "Consent Form," Harvard Medical School, accessed January 5, 2015, http://www.personalgenomes.org/static/docs/harvard/PGP_Consent_Approved_02212013.pdf.
70 See Johnson, "George Church, Harvard Genetics Professor, Banters with Stephen Colbert."
71 Interview with author, May 29, 2012.
72 Interview with author, July 24, 2013.
73 Lunshof et al., "Personal Genomes in Progress: From the Human Genome Project to the Personal Genome Project," 47–60.
74 Bolnick et al., "The Science and Business of Genetic Ancestry Testing"; Manrai et al., "Genetic Misdiagnoses and the Potential for Health Disparities," 655–65.
75 For a discussion of Georgia Dunston's work, see chapter 2.
76 "About PGP Harvard," Personal Human Genome Project: Harvard, accessed January 5, 2015, http://www.personalgenomes.org/harvard/about.
77 "George M. Church's Tech Transfer, Advisory Roles, and Funding Sources," Harvard Molecular Technologies, accessed January 5, 2015, http://arep.med.harvard.edu/gmc/tech.html.

78 Ibid.

79 Yet it should be noted that even John Sulston believed companies played an important role in sequencing the human genome. Indeed, Sulston dubbed Merck a "white knight" when it saved expressed sequence tag data from enclosure by Venter and his corporate backers. Sulston and Ferry, *Common Thread: A Story of Science, Politics, Ethics, and the Human Genome*, 139.

80 For historical accounts of the construction of public performances as central to science, see Ezrahi, *Descent of Icarus*; Shapin and Schaffer, *Leviathan and the Air-Pump: Hobbes, Boyle, and the Experimental Life*.

81 Interview with author, July 25, 2013.

82 Foner, *Story of American Freedom*, 9.

83 As with other liberal democratic rights, at the founding of the United States and for centuries afterward, this right was granted to the elite few who counted as citizens.

84 Tennessee Supreme Court quoted in Foner, *Story of American Freedom*, 53. See text of original case in Humphreys, *Reports of Cases Argued and Determined in the Supreme Court of Tennessee*, vol. 4.

85 Arendt reminds us that in its original usage, *private* referred to the privative state. It literally meant a state of being deprived. Humans who existed only in the private realm were not fully human. Arendt, *Human Condition*, 38.

86 Boyle, *Shamans, Software, and Spleens: Law and the Construction of the Information Society*.

87 Nussbaum, "Say Everything: The Future Belongs to the Uninhibited."

88 Arendt, *Human Condition*, 28.

89 Staudte was the premier film director in East Germany in the immediate postwar period. For a critical appraisal and description of *Rotation*, see "Rotation," East German Cinema Blog, accessed August 17, 2016, https://eastgermancinema.com/2011/03/07/rotation/.

90 Arendt, *Human Condition*, 31.

91 One way to understand the difference is that it is similar to the one Michel Foucault observed between states of domination and powers of freedom. Foucault argued that we need to distinguish between "the relationships of power as strategic games between liberties" and "states of domination, which are what we ordinarily call power." Foucault, "'Omnes et Singulatim': Toward a Critique of Political Reason," 299. Arendt and Staudte worried about the states of domination. Social media exemplifies relationships of power as strategic games between liberties.

92 See "Consent Form," Harvard Medical School, accessed January 26, 2015, http://www.personalgenomes.org/static/docs/harvard/PGP_Consent_Approved_02212013.pdf.

93 PGP participants' interests in their own health is also evident in the discussions at the GET conferences. Author field notes.

94 Interview with author, July 16, 2012.

95 See Chu, "MeForYou Campaign Rallies Public to Join Push for Precision Medicine."

96 Translated into English this phrase means, "The common good before the private good." The phrase was stamped on Reichsmark coins from 1933.

97 Meyer, "Reflections of a Re-Identification Target, Part 1."

98 Angrist, "I Never Promised You a Walled Garden."

99 Meyer, "Ethical Concerns, Conduct, and Public Policy for Re-Identification and De-Identification Practices."

100 Ibid.

101 Lunshof et al., "From Genetic Privacy to Open Consent," 5.

102 Consistent with their belief that transparency is the key to addressing ethical issues, the PGP response to the re-identification debate was to allow people to opt into providing their names. This would make clear who was truly comfortable with being identified. See Connelly, "PGP Harvard Updates—Including a New 'Real Name' Option."

103 Meyer, "Reflections of A Re-Identification Target, Part 1."

104 See https://wellcome.ac.uk/press-release/research-funders-outline-steps-prevent-re-identification-anonymised-study-participants, accessed February 10, 2017.

105 Hayden, "Taking as Giving: Bioscience, Exchange, and the Politics of Benefit-Sharing," 751.

106 For discussions of how rights to privacy and property might be reworked to serve a more just collective society, see Thompson, *Good Science*, 181–88. See also Harrison, "Neither Moore nor the Market: Alternative Models for Compensation Contributors of Human Tissue," 77–105. See also Laurie, *Genetic Privacy: A Challenge to Medico-Legal Norms*.

107 Arendt, *Between Past and Future*, 236–37.

108 It also did not bring attention to the ways in which PGP leaders worked to craft a project that respected the wishes of its participants—for example, whether they wanted their names associated with their data. See Connelly, "PGP Harvard Updates—Including A New 'Real Name' Option."

109 Dean, *Democracy and Other Neoliberal Fantasies*; Tkacz, *Wikipedia and the Politics of Openness*.

110 Here I draw on Charis Thompson's formulation of "good science" as a science whose ethical dimensions cannot be denied. Minding that relation leads to both more robust science and a greater understanding and mitigation of social injustices. Thompson, *Good Science*, 28.

111 In an interview with the *Los Angeles Times*, Church reportedly argued that we should "make it so that people don't care about it [privacy]." Brown, "Geneticist on DNA Privacy: Make It So People Don't Care."

112 Barad, *Meeting the Universe Halfway: Quantum Physics and the Entanglement of Matter and Meaning*.

CHAPTER EIGHT

1 Klein, *Shock Doctrine: The Rise of Disaster Capitalism.*

2 When Facebook went "public" in 2012, millions became millionaires overnight, causing housing prices to double and rents to soar in San Francisco, forcing out many longtime residents. See Ludka, "Meet the New Facebook Millionares."

3 Richtel, "Battle of Mission Bay," BU 1, 5.

4 Said, "Pfizer to Open Research Center in SF's Mission Bay."

5 Kneebone, Nadeau, and Berube, "Re-Emergence of Concentrated Poverty: Metropolitan Trends in the 2000s."

6 Hunters Point Family, "Our History," http://hunterspointfamily.org/who-we-are/our-history/.

7 Barros, "UCSF Plans to Shutter Clinic Serving Minority Youth."

8 While this book focuses on the question of the meaning of genomics for human beings, as Donna Haraway reminds us, the well-being of all beings is entangled. Questions about what genomics means for conservation and responding to a damaged planet today also are pressing questions. For one site of this discussion, see the recent debate over de-extinction. See Gross, "De-Extinction Debate: Should Extinct Species Be Received?"; Haraway, *When Species Meet.*

9 Interview with author, May 1, 2000.

10 But also $115 million more support from financial backers. See "23andMe Raises $115M in Series E Financing Round." *Genome Web*, October 14, 2015. https://www.genomeweb.com/business-news/23andme-raises-115m-series-e-financing-round, accessed October 23, 2015.

11 Brown, *Undoing the Demos: Neoliberalism's Stealth Revolution*, 128.

12 Angrist, *Here Is a Human Being: At the Dawn of Personal Genomics.* However, as Angrist explained to me in January 2016, Illumina now gives its customers iPads with their sequences. More generally, data now live in "the cloud." One needs Internet access and permission codes, not a powerful computer, to see their genome sequence.

13 Over the course of the decade, more than once I heard a genome scientist suggest that thinking about ethical issues wasted time that could be used to save lives.

14 Arendt found strength for her arguments in the writing of the Greeks and Romans, medieval theologians, and Enlightenment thinkers—from Aristotle to Thomas Aquinas to Adam Smith.

15 This publicity comes from people gathering around a common thing and seeing it for all of its "utter diversity." Seeing through difference—not through the conformism of an administrated mass society—generates shared meaning. Arendt, *Between Past and Future*, 55–57.

16 This, however, may be fast changing. The opening days of the Trump presidency witnessed a dramatic removal of previously available public data

from federal websites. See Berman, "We Rely on the Government for Lots of Data."

17 Obama, "Remarks of the President in Welcoming Senior Staff and Cabinet Secretaries."

18 Ehley, "Why Obama's $30B Digital Health Record Plan Is Failing."

19 That digitization of medical records became a focus of the Obama administration highlights the underlying complexities. Medical records are considered by many to be private information. Who should see them and for what ends is far from a settled matter.

20 For a more extended consideration of the ways in which informatic capitalism has unmoored publicity as a norm of liberal democracies, see Dean, *Publicity's Secret: How Technoculture Capitalizes on Democracy*.

21 For an effort to overturn the Myriad Genetics monopoly control of this information on *BRCA1* and *BRCA2* variants, see Free the Data, http://www.free-the-data.org/, accessed August 24, 2016.

22 McCormick et al., "Giving Office-Based Physicians Electronic Access to Patients' Prior Imaging and Lab Results Did Not Deter Ordering of Tests," 488–96.

23 As I finished writing this book, the arrival of the Trump administration drew these complexities of the value of open data to the fore. People from around the world worked to ensure some data remained publicly available (for example, climate data) while working to remove other data (for example, library user data that might be used for government surveillance). See Thielman, "Libraries Promise to Destroy User Data to Avoid Threat of Government Surveillance"; Brown, "A Coalition of Scientists Keeps Watch on the U.S. Government's Climate Data."

24 Allen, "Biden's Cancer Bid Exposes Rift among Researchers."

25 In this sense, genomics bears out what social theorists have observed in many other domains. The contemporary period is defined by a shift from liberal democratic government to neoliberal governance. This shift is marked by the "economization" of all areas of life. Brown, *Undoing the Demos: Neoliberalism's Stealth Revolution*, 122.

26 See "Chapter 1: Welcome and Corporate Overview," https://www.youtube.com/watch?v=fAChUEBlnrM, accessed September 16, 2015.

27 However, Coriell still sends out newsletters informing HapMap communities about how their samples are being used. Recently the NIH decided to retire the HapMap resource, citing security concerns. See "NCBI Retiring HapMap Resource," June 16, 2016, accessed August 24, 2016, http://www.ncbi.nlm.nih.gov/variation/news/NCBI_retiring_HapMap/.

28 Munro, "Class Action Law Suit Filed against 23andme."

29 Birch, "Rethinking Value in the Bio-Economy: Finance, Assetization, and the Management of Value," 1–31.

30 Indeed, I am a member of the UCSF gym in Mission Bay, which I called the Gattaca gym long before I had thought very deeply about the transformations happening around me. Its spaces are grand and futuristic.

31 "Want to help heal the world? Start by sharing your health data," TEDMED, May 13, 2013, accessed September 21, 2015, http://blog.tedmed.com/tag/meforyou/.

32 Kim, "UCSF Wins Four CASE Awards of Excellence."

33 Dean, *Publicity's Secret: How Technoculture Capitalizes on Democracy*, 14.

34 Jaffe, "End in Sight for Revision of Us Medical Research Rules," 1225–26.

35 However, on June 29, 2016, the National Academy of Sciences issued a report challenging the wisdom of this call for consent for secondary uses of de-identified biospecimens. Indeed, they called for the executive branch to recall its entire Notice for Proposed Rulemaking for the Common Rule. See "Congress Should Create Commission to Examine the Protection of Human Participants in Research; Notice of Proposed Rulemaking to Revise Common Rule Should Be Withdrawn," *National Academies of Sciences, Engineering, Medicine*, June 29, 2016, accessed August 24, 2016, http://www8.nationalacademies.org/onpinews/newsitem.aspx?RecordID=21824.

36 See, for example, the conclusion of Kimberly TallBear's book, *Native American DNA: Tribal Belonging and the False Promise of Genetic Science*. I also describe some of these efforts in chapter 6 my book *Race to the Finish*.

37 See "Federal Policy for the Protection of Human Subjects," Federal Register, accessed October 20, 2015, https://www.federalregister.gov/articles/2015/09/08/2015-21756/federal-policy-for-the-protection-of-human-subjects#p-243.

38 Charis Thompson and Ruha Benjamin observed similar citizen concerns about the state of California's investments in stem cell research. See Thompson, *Good Science*; and Benjamin, *People's Science: Bodies and Rights on the Stem Cell Frontier*. The Greenlining Institute in Oakland explicitly addressed these issues at its Towards Fair Cures conference in 2006. See the issue brief on diversity in stem cell research, Tayag, "Toward Fair Cures: Diversity Policies in Stem Cell Research."

39 I want to thank Donna Haraway for sharing this formulation of knowledge as a problem of energy as we walked along the San Francisco Bay off Third Street.

40 Venter, *A Life Decoded*, 250–51.

41 Author field notes, April 13, 2011.

42 Hayden, "Genome Researchers Raise Alarm over Big Data."

43 Vaugham, "How Viral Cat Videos Are Warming the Planet."

44 See "Welcome," H3Africa: Human Heredity and Health in Africa, accessed October 6, 2015, h3africa.org.

45 These drugs can cost as much as ten to twenty thousand dollars a month.

See the discussion of whether this is just or not at the UCSC Just Data meeting in May 2016 at Karuna Jaggar, Twitter post, May 19, 2016, 12:15 p.m., https://twitter.com/karunajaggar/status/733375465164054530.

46 I thank Steve Sturdy for pointing out that ten out of the eleven most expensive drugs the National Health Service in England prescribes are produced through biotechnology, and two are tyrosine kinase inhibitors, one of the only novel form of a drug to arise from genomics research. See "Hospital Prescribing England 2013–14," Health and Social Care Information Centre, 2014, https://www.digital.nhs.uk/catalogue/PUB15883/hosp-pres-eng-201314-un-dat.xlsx.

47 Koenig, "Have We Asked Too Much of Consent?," 33–34.

48 I thank the Max Planck Institute for the History of Science Arts of Judgment Working Group for inspiring conversations about what we might mean by the arts of judgment.

49 These are important subjects of legal and science studies. See, for example, in science studies, Galison, *How Experiments End*.

50 To remind the reader, here I use Arendt's formulation of inter-est to denote that which lies between us.

51 Rawls, *Theory of Justice*, 3.

52 Clarke et al., "Biomedicalization: Technoscientific Transformations of Health, Illness, and U.S. Biomedicine," 161–94.

53 See chapter 2 and Sulston and Ferry, *Common Thread: A Story of Science, Politics, Ethics, and the Human Genome*.

54 Wade, "A Decade Later, Genetic Map Yields Few New Cures."

55 See Fleck et al., "Summary Dialogue Report"; Bonham et al., "Community-Based Dialogue: Engaging Communities of Color in the United States' Genetics Policy Conversation," 325–59.

56 Arendt, *Human Condition*, 26.

57 Ibid., 180.

58 Arendt, *Between Past and Future*, 218.

59 Arendt, *Human Condition*, 5.

60 For a social media platform discussion of the problem of the echo chamber, see Jones, "How Social Media Created an Echo Chamber for Ideas." For a recent rebuttal from Facebook, see Kokalitcheva, "Is Facebook a Political Echo Chamber? Social Network Says No."

61 Arendt, *Human Condition*, 7.

62 See "Office of the Press Secretary," NIH, accessed October 8, 2015, https://www.genome.gov/10001356.

63 Indeed, the BBC describes its effort "to curate a comprehensive history of every radio and TV programme ever broadcast by the corporation, and make that available to the public" as its Genomics Initiative. I would like to thank

Jenny Bangham for bringing this to my attention. See "Genome-Radio Times Archive Now Live," BBC, accessed October 9, 2015, http://www.bbc.co.uk/blogs/aboutthebbc/entries/108fa5e5-cc28-3ea8-b4a0-129912a74efc.

64 See "Precision Medicine Initiative Participant Engagement and Health Equity Workshop (Day 1)," NIH Video Casting and Podcasting, http://videocast.nih.gov/summary.asp?Live=16498&bhcp=1.

65 See "The Precision Medicine Initiative Cohort Program—Building a Research Foundation for 21st Century Medicine."

66 See Patil and Devaney, "Next Steps in Developing the Precision Medicine Initiative."

67 See Graber, "The Problem with Precision Medicine"; Kroll, "Obama's Precision Medicine Initiative: Paying for Precision Drugs Is the Challenge."

68 Putnam, "Bowling Alone: America's Declining Social Capital," 65–78.

69 Durkheim, *Suicide*.

70 See "Redefining Possible," UCSF, accessed August 24, 2016, http://www.ucsfredefiningpossible.com/.

71 See http://www.meforyou.org/, accessed August 24, 2016.

72 For a description of this initiative, see "Treehouse Childhood Initiative," UCSC, accessed August 24, 2016, https://treehouse.soe.ucsc.edu/.

73 Roy was particularly concerned about the difficulty of imagining worlds outside of capitalism. See Roy, "Can We Leave the Bauxite in the Mountain? Field Notes on Democracy." I want to thank Ruha Benjamin for calling my attention to this speech, and for her analysis of this problem in her book, *People's Science*.

74 Arendt, *Human Condition*, 97, 173, 176, 184.

75 Ibid., 182.

76 I can only begin in this conclusion to open up our imaginaries of genomics to consider the much-more-than-human lives that are stake. For help, I turn to Donna Haraway, who in her formulation of "kin and kind" uses Shakespearean language to unsettle our notions of how we build relationships. Kin, as family, may not be the kindest. Kin, as care, does not come only from cultivating relations with human beings. "Kin," Haraway writes, "is an assembling sort of word. All critters share a common 'flesh,' laterally, semiotically, and genealogically." Haraway, "Anthropocene, Capitalocene, Plantationocene, Chthulucene: Making Kin," 159–65. Arendt also understood the importance of these earthly connections. "The earth," she wrote, "is the very quintessence of the human condition." Arendt, *Human Condition*, 2.

77 NCI grant number U54 CA15350605. For information about this grant, see https://projectreporter.nih.gov/project_info_description.cfm?aid=8729265&icde=31790021&ddparam=&ddvalue=&ddsub=&cr=1&csb=default&cs=ASC, accessed November 5, 2016.

78 For a description of the original call, see http://grants.nih.gov/grants/guide/rfa-files/rfa-ca-09-032.html. For a description of the "parent" grant see: https://gsspubssl.nci.nih.gov/nciportfolio/search/details%3Bjsessionid=D7AF989619BAD64244F4A073B738C86C?action=abstract&grantNum=3U54CA153506-03S2&grantID=8534909&grtSCDC=FY%202012&absID=8537579&absSCDC=CURRENT.

79 The quotes in this section come from reading group discussions conducted with the support of an administrative supplement to the Alameda County Network Program for Reducing Cancer Disparities grant. NCI grant number U54 CA15350605. I would like to thank Julie Harris-Wai and Jaennika Aniag for their help understanding this research and its outcomes.

80 Ibid.

81 Ibid.

82 It is important to qualify which stories I think help us, and which ones I think do not. The difference between talk and speech here is crucial. Stories that function to serve instrumental ends—for example, MeForYou's story about Georgia designed to pull at heart strings in order to convince citizens to give their tissue and data to UCSF—are explicitly not what we need.

83 King, *Interview with Patricia King: Belmont Oral History Project.*

84 For this formulation of justice, see my discussion of Rawls in the introduction.

85 For a discussion of our approach, and examples of collective projects, see Science and Justice Research Center (Collaborations Group), "Experiments in Collaboration: Interdisciplinary Graduate Education in Science and Justice," 1–5. Reardon et al., "Science and Justice: The Trouble and the Promise."

86 For more details on these examples, see Reardon et al., "Science and Justice: The Trouble and the Promise."

87 Our work was generously supported by the National Science Foundation under grant no. SES-0933027. Any opinions, findings, and conclusions or recommendations expressed in this work do not necessarily reflect the views of the NSF.

88 Bonham et al., "Community-Based Dialogue: Engaging Communities of Color in the United States' Genetics Policy Conversation," 5–6.

89 Reardon, *Race to the Finish: Identity and Governance in an Age of Genomics.*

90 Lambro, *Fat City: How Washington Wastes Your Taxes.*

91 To be successful, any such effort should learn from the shortcomings of the OTA and build an agency that is less bound by the politics of the moment and that is more responsive to broad public concerns. For an assessment of the OTA, see Jasanoff, *The Ethics of Invention.*

92 Callon, "Elements of a Sociology of Translation: Domestication of the Scallops and the Fishermen of St Brieuc Bay," 196–233.

93 Interview with author, November 16, 2010.

94 Interview with author, February 18, 2011.

95 Ibid.

96 It also required humans who could "read" data that computers still struggled to comprehend (such as data contained in graphs and charts).

97 Interview with author, April 14, 2010.

98 Stevens, *Life out of Sequence: A Data-Driven History of Bioinformatics.*

99 Thank you to Steve Sturdy for suggesting the formulation "thinking-in-vivo."

100 For the technical changes that made blood donation possible (freezer, etc.) see Bangham, "Blood Groups and the Rise of Human Genetics in Mid-Twentieth Century Britain."

101 Pemberton, *Bleeding Disease: Hemophilia and the Unintended Consequences of Medical Progress.*

102 Waldby and Mitchell, *Tissue Economies: Blood, Organ, and Cell Lines in Late Capitalism*, 2. On the continued significance of blood donation as a voluntary, altruistic act, see also Tutton, "Person, Property, and Gift: Exploring Languages of Tissue Donation to Biomedical Research," 17–38.

103 Waldby and Mitchell, *Tissue Economies: Blood, Organ and Cell Lines in Late Capitalism*, 22. For a historical account of the challenges to altruism created by the transformation of blood donation from direct body-to-body exchange to an indirect system of collection mediated by a bureaucracy, see Whitfield, *Who Is My Stranger?*

104 Waldby and Mitchell, *Tissue Economies: Blood, Organ and Cell Lines in Late Capitalism*, 24.

105 A decade ago, science studies scholar Sheila Jasanoff raised a similar question about the viability of the idea of a social contract with science in an age where wealth creation and knowledge creation were intimately entangled. Jasanoff, *Designs on Nature: Science and Democracy in Europe and the United States.*

106 For a discussion of the closure of the US National Center for Biotechnology Information Sequence Read Archive, see Genome Biology Editorial Team, "Closure of the NCBI SRA and Implications for the Long-Term Future of Genomics and Data Storage." For the opening of DNAnexus, a bioinformatics company designed to address the problem of burgeoning raw genomic data, see DNAnexus, "DNAnexus Launches Web-Based Serviced to Ease Data Bottleneck in Next-Generation DNA Sequencing."

EPILOGUE

1 Thies, *Hitler's Plans for Global Domination: Nazi Architecture and Ultimate War Aims*, 87.

2 Tempelhofer Unfreiheit, "Forced Labor in Berlin and Brandenburg," http://www.tempelhofer-unfreiheit.de/en/forced-labor-berlin-brandenburg/, accessed November 16, 2016.

3 Fahey, "How Berliners Refused to Give Tempelhof Airport over to Developers."

4 For a discussion of the problem of housing in Berlin and its relationship to the Tempelhof vote, see Silke, "Berlin Voters Claim Tempelhof."

5 Poschardt quoted in Fahey, "How Berliners Refused to Give Tempelhof Airport over to Developers."

6 See Knight, "Berlin to Build on Tempelhof Despite Drop in Refugees."

7 See, for example, "Welcome Dinner Briefly Explained," Welcome Dinner Berlin, accessed August 24, 2016, www.welcomedinnerberlin.de; and "Give Something Back to Berlin," givesomethingbacktoberlin.com, accessed August 24, 2016.

8 Borry et al., "Legislation on Direct-to-Consumer Genetic Testing in Seven European Countries," 715–21.

9 Arendt, *Eichmann in Jerusalem: A Report on the Banality of Evil*. Even if one accepts the bureaucracy argument, there are additional questions about what kind of a state could have engendered it.

10 Kühl, *Nazi Connection: Eugenics, American Racism, and German National Socialism*.

11 Kuwait, for example, recently passed a law requiring all citizens, residents, and visitors to provide a DNA sample in order to enter or stay in the country. See Rivero, "Kuwait's New DNA Collection Law Is Scarier Than We Ever Imagined." Many nations around the world have growing DNA forensic databases.

ACKNOWLEDGMENTS

1 Haraway, *Simians, Cyborgs, and Women*, 196.

2 Before settling on *The Human Condition*, Arendt considered the title *Amor Mundi* or *Love of the World*. She believed it of critical importance to love the world even in the face of horrific actions and tremendous suffering. Engaging all that is—neither uncritically accepting or celebrating—fosters thought and love: passionate thinking.

3 For a description of the Science and Justice Training Program and the students who have taken part, see https://scijust.ucsc.edu/training/, accessed November 1, 2016.

4 Reardon, "Democratic Mis-Haps"; Reardon, "Genomics' Problem of Communication"; Reardon, "The 'Persons' and 'Genomics' of Personal Genomics."

BIBLIOGRAPHY

Adams, Vincanne, Michelle Murphy, and Adele E. Clarke. "Anticipation: Technoscience, Life, Affect, Temporality." *Subjectivity* 28 (2009): 246–65.

Allen, Arthur. "Biden's Cancer Bid Exposes Rift among Researchers." *Politico*, January 31, 2016. http://www.politico.com/story/2016/01/joe-biden-cancer-researchers-rift-218465.

Anderson, Chris. *Free: The Future of a Radical Price*. New York: Hyperion, 2009.

Anderson, Garnet L., Howard L. Judd, Andrew M. Kaunitz, David H. Barad, Shirley A. A. Beresford, Mary Pettinger, James Liu, et al. "Effects of Estrogen Plus Progestin on Gynecologic Cancers and Associated Diagnostic Procedures: The Women's Health Initiative Randomized Trial." *JAMA* 290 (2003): 1739–48.

Andrews, Travis, M. "Twitter Suspends Prominent Alt-Right Accounts." *Los Angeles Times*, November 16, 2016. http://www.latimes.com/business/technology/la-fi-tn-twitter-alt-right-20161116-story.html.

Angier, Natalie. "Do Races Differ? Not Really, Genes Show." *New York Times*, August 22, 2000. http://www.nytimes.com/2000/08/22/science/do-races-differ-not-really-genes-show.html?pagewanted=all.

———. "Scientist at Work: Mary-Claire King; Quest for Genes and Lost Children." *New York Times*, April 27, 1993. http://www.nytimes.com/1993/04/27/science/scientist-at-work-mary-claire-king-quest-for-genes-and-lost-children.html?pagewanted=all.

Angrist, Misha. "Eyes Wide Open: The Personal Genome Project, Citizen Science, and Veracity in Informed Consent." *Personal Medicine* 6, no. 6 (2009): 691–99.

———. *Here Is a Human Being: At the Dawn of Personal Genomics*. New York: Harper Collins, 2010.

———. "I Never Promised You a Walled Garden." *Bill of Health Blog*. http://blogs.law.harvard.edu/billofhealth/2013/05/25/i-never-promised-you-a-walled-garden-re-identification-symposium/.

Ankeny, Rachel A. "The Natural History of *Caenorhabditis elegans* Research." *Nature Reviews Genetics* 2, no. 6 (2001): 474–79.

Appiah, Anthony. *Cosmopolitanism: Ethics in a World of Strangers*. Princeton, NJ: Princeton University Press, 2006.

Arendt, Hannah. *Between Past and Future*. New York: Penguin, 2006.

———. *Eichmann in Jerusalem: A Report on the Banality of Evil*. New York: Penguin, 2006.

———. *The Human Condition*. Chicago: University of Chicago Press, 1998.

———. *On Violence*. New York: Harcourt, 1969.

———. *The Origins of Totalitarianism*. New York: Harcourt Brace, 1979.

Ashburner, Michael. *Won for All: How the* Drosophila *Genome Was Sequenced*. Cold Spring Harbor, NY: Cold Spring Harbor Laboratory Press, 2006.

Avey, Linda. "Inspiration Comes in Many Forms." *The Life and Times of Lily Mendel* (blog), November 4, 2009. http://lillymendel.blogspot.com/2009/11/inspiration-comes-in-many-forms-idea-to.html.

———. "Reminiscing." *The Life and Times of Lily Mendel* (blog), June 23, 2010. http://lillymendel.blogspot.com/2009/11/inspiration-comes-in-many-forms-idea-to.html.

Axelrod, Jim. "The Immortal Henrietta Lacks." *CBS News*, March 15, 2010. Accessed June 7, 2013. http://www.cbsnews.com/8301–3445_162–6300824/the-immortal-henrietta-lacks/.

Baker, Dean. "The Housing Crash Recession: How Did We Get Here?" *PBS Now*, March 21, 2008. www.pbs.org/now/shows/412/housing-recession.html.

Baldwin, James. *The Evidence of Things Not Seen*. New York: Holt, Rinehart and Winston, 1985.

Baldwin, Peter. *The Copyright Wars: Three Centuries of Trans-Atlantic Battle*. Princeton, NJ: Princeton University Press, 2014.

Ball, Madeleine, Jason Bobe, Michael Chou, Tom Clegg, Preston Estep, Jeantine Lunshof, Ward Vandewege, et al. "Harvard Personal Genome Project: Lessons from Participatory Public Research." *Genome Medicine* 6, no. 2 (2014): 10.

Bangham, Jenny. "Blood Groups and the Rise of Human Genetics in Mid-Twentieth Century Britain." Max Planck Institute for the History of Science. https://www.mpiwg-berlin.mpg.de/en/research/projects/jbangham_blood_groups.

Barad, Karen. *Meeting the Universe Halfway: Quantum Physics and the Entanglement of Matter and Meaning*. Durham, NC: Duke University Press, 2007.

Barros, Joe Rivano. "UCSF Plans to Shutter Clinic Serving Minority Youth." *Mission Local*, March 15, 2016. http://missionlocal.org/2016/03/ucsf-plans-to-shutter-clinic-serving-minority-youth/.

Bartlick, Silke. "Berlin Voters Claim Tempelhof." *DW*, May 27, 2014. Accessed September 13, 2016. http://www.dw.com/en/berlin-voters-claim-tempelhof/a-17663944.

Baudrillard, Jean. *The Ecstasy of Communication*. New York: Semiotext(e), 1988.

Beeson, Diane, and Troy Duster. "African-American Perspectives on Genetic Testing." In *The Double-Edged Helix: Social Implications of Genetics in a Diverse Society*, edited by Joseph S. Alper, Catherine Ard, Adrienne Asch, Jon Beckwith, Peter Conrad, and Lisa N. Geller, 151–74. Baltimore: Johns Hopkins University Press, 2002.

Benjamin, Ruha. "A Lab of Their Own: Genomic Sovereignty as Postcolonial Science Policy." *Policy and Society* 28, no. 4 (2009): 341–55.

———. *People's Science: Bodies and Rights on the Stem Cell Frontier*. Stanford, CA: Stanford University Press, 2013.

Bentham, Jeremy. "Of Publicity." In *The Works of Jeremy Bentham*, vol. 2, edited by John Bowring, 310. New York: Russell and Russell, 1962.

Berlant, Lauren. *Cruel Optimism*. Durham, NC: Duke University Press. 2011.

Berman, Mark. "We Rely on the Government for Lots of Data: What Happens to That in the Trump Era of 'Alternative Facts' "? *Washington Post*, January 23, 2017. Accessed February 10, 2017. https://www.washingtonpost.com/news/post-nation/wp/2017/01/23/we-rely-on-the-government-for-lots-of-data-what-happens-to-that-in-the-era-of-alternative-facts/?utm_term=.7fd84d89ce7d.

Birch, Kean. "Rethinking Value in the Bio-Economy: Finance, Assetization, and the Management of Value." *Science, Technology, and Human Values* (August 2016): 1–31.

Bird, Kai, and Martin J. Sherwin. *American Prometheus: The Triumph and Tragedy of J. Robert Oppenheimer*. New York: A. A. Knopf, 2005.

Blakey, Michael. "Beyond European Enlightenment: Toward a Critical and Humanistic Biology." In *Building a New Biocultural Synthesis: Political-Economic Perspectives on Human Biology*, edited by Alan H. Goodman and Thomas L. Leatherman, 379–406. Ann Arbor: University of Michigan Press, 1998.

Bliss, Catherine. *Race Decoded: The Genomic Fight for Social Justice*. Palo Alto, CA: Stanford University Press, 2012.

Bobe, Jason. "Open Humans Network: Find Equitable Research Studies and Donate Your Personal Data for Public Benefit." *NewsChallenge.org*. Last modified September 17, 2013. https://www.newschallenge.org/challenge/healthdata/evaluation/open-humans-network-find-equitable-research-studies-and-donate-your-personal-health-data-for-public-benefit.

Bolnick, Deborah A., Duana Fullwiley, Troy Duster, Richard S. Cooper, Joan H. Fujimura, Jonathan Kahn, Jay S. Kaufman, et al. "The Science and Business of Genetic Ancestry Testing." *Science* 318, no. 5849 (2007): 399–400. doi: 10.1126/science.1150098.

Bonham, Vence L., Toby Citrin, Stephen M. Modell, Tené Hamilton Franklin, Esther W. B. Bleicher, and Leonard M. Fleck. "Community-Based Dialogue: Engaging Communities of Color in the United States' Genetics Policy Conversation." *Journal of Health Politics, Law, and Policy* 34, no. 3 (2009): 325–59.

Borry, Pascal, Rachel E. van Hellemondt, Dominique Sprumont, Camilla Fittipaldi Duarte Jales, Emmanuelle Rial-Sebbag, Tade Matthias Spranger, Liam Curren, et al. "Legislation on Direct-to-Consumer Genetic Testing in Seven European Countries." *European Journal of Human Genetics* 20, no. 7 (2012): 715–21.

Boseley, Sarah, and Helen Carter. “He Stripped the Organs from Every Dead Child He Touched.” *The Guardian*, January 31, 2001. https://www.theguardian.com/society/2001/jan/31/health.alderhey.

Boston Women’s Health Book Collective. *Our Bodies, Ourselves: A New Edition for a New Era*. New York: Scribner, 2005.

Bourdieu, Pierre. *Distinction: A Social Critique of the Judgement of Taste*. New York: Routledge, 1986.

Bowen, Anthony, and Arturo Casadevall. “Increasing Disparities between Resource Inputs and Outcomes, as Measured by Certain Health Deliverables, in Biomedical Research.” *Proceedings of the National Academy of Sciences* 112, no. 36 (2015). doi: 10.1073/pnas.1504955112.

Bowker, Geoffrey C., and Susan Leigh Star. *Sorting Things Out: Classification and Its Consequences*. Cambridge, MA: MIT Press, 1999.

Boyle, James. *Shamans, Software, and Spleens: Law and the Construction of the Information Society*. Cambridge, MA: Harvard University Press, 1996.

Brainard, Curtis. “HeLa-Cious Coverage: Media Overlook Ethical Angles of Henrietta Lacks Story.” *Columbia Journalism Review* (March 28, 2013).

Brockman, John. *Digerati: Encounters with the Cyber Elite*. San Francisco: Hardwired, 1996.

Brower, Vicki. “A Second Chance for Hormone Replacement Therapy?” *EMBO Reports* 4, no. 12 (2003): 1112–15.

Brown, Alleen. “A Coalition of Scientists Keeps Watch on the U.S. Government’s Climate Data.” *Intercept*, January 27, 2017. https://theintercept.com/2017/01/27/a-coalition-of-scientists-keeps-watch-on-the-u-s-governments-climate-data/.

Brown, Eryn. “Geneticist on DNA Privacy: Make It So People Don’t Care.” *Los Angeles Times*, January 18, 2013.

Brown, Wendy. *Undoing the Demos: Neoliberalism’s Stealth Revolution*. New York: Zone Books, 2015.

Busby, Helen, and Paul Martin. “Biobanks, National Identity, and Imagined Communities: The Case of UK Biobank.” *Science as Culture* 15, no. 3 (2006): 237–51.

Butler, Judith. *Giving an Account of Oneself*. New York: Fordham University Press, 2005.

———. *Precarious Life: The Powers of Mourning and Violence*. London: Verso, 2004.

Callon, Michel. “Elements of a Sociology of Translation: Domestication of the Scallops and the Fishermen of St Brieuc Bay.” In *Power, Action, and Belief: A New Sociology of Knowledge?*, edited by John Law, 196–233. London: Routledge, 1986.

Cann, Rebecca, Mark Stoneking, and Alan C. Wilson. “Mitochondrial DNA and Human Evolution.” *Nature* 325 (1987): 31–36.

Castells, Manuel. *The Rise of the Network Society*. Oxford: Oxford University Press, 2000.

Cavalli-Sforza, Luca. "The Human Genome Diversity Project." In *An Address Delivered to a Special Meeting of UNESCO*. Paris, France, 1994.

———. "Race Differences: Genetic Evidence." In *Plain Talk about the Human Genome Project: A Tuskegee University Conference on Its Promise and Perils and Matters of Race*, edited by by Edward Smith and Walter Sapp, 51–58. Tuskegee, AL: Tuskegee University Press, 1997.

Cavalli-Sforza, Luca, Alan C. Wilson, Charlie Cantor, Robert M. Cook-Deegan, and Mary-Claire King. "Call for a Worldwide Survey of Human Genetic Diversity: A Vanishing Opportunity for the Human Genome Project." *Genomics* 11 (Summer 1991): 490–91.

Chief Scientist Office. *Guidance for Applicants to the Genetics and Health Care Initiative*. Edinburgh: Scottish Executive, 2003.

Chu, Louise. "MeForYou Campaign Rallies Public to Join Push for Precision Medicine." *UCSF News Center*, May 3, 2013. Accessed August 16, 2015. http://www.ucsf.edu/news/2013/05/105616/meforyou-campaign-rallies-public-join-push-precision-medicine.

Church, George. "The Personal Genome Project." *Molecular Systems Biology* 1, no. 1 (2005).

Church, George M. "Genomes for All." *Scientific American* (2006): 47–54.

———. "Human Genome Project (HGP) History (a Personal Account)." Accessed October 16, 2016. http://arep.med.harvard.edu/gmc/HGP.html.

Church, George M., and Ed Regis. *Regenesis: How Synthetic Biology Will Reinvent Nature and Ourselves*. New York: Basic Books, 2012.

Clarke, Adele, Jennifer Fishman, Jennifer Fosket, Laura Mamo, and Janet Shim. "Biomedicalization: Technoscientific Transformations of Health, Illness, and U.S. Biomedicine." *American Sociological Review* 68 (2003): 161–94.

Clarke, Roger. "Information Wants to Be Free . . ." *Roger Clarke's Web-Site*, 2000; August 12, 2015. http://www.rogerclarke.com/II/IWtbF.html.

Clinton, William J. "Statement Transcribed on Decoding of Genome." *New York Times*, June 27, 2000.

Cohen, Noam. "Egyptians Were Unplugged, and Uncowed." *New York Times*, February 20, 2011. http://www.nytimes.com/2011/02/21/business/media/21link.html?_r=0.

Collins, Francis S. "Contemplating the End of the Beginning." *Genome Research* 11 (2001): 641–43.

———. *The Language of God: A Scientists Presents Evidence for Belief*. New York: Free Press, 2006.

Collins, Francis S., Lisa D. Brooks, and Aravinda Chakravarti. "A DNA Polymorphism Discovery Resource for Research on Human Genetic Variation." *Genome Research* 8 (1998): 1229–31.

Connelly, Abram. "PGP Harvard Updates—Including a New 'Real Name' Option." *News and Updates about the Personal Genome Project* (December 18, 2014).

Cook-Deegan, Robert M. "The Alta Summit, December 1984." *Genomics* 5 (October 1989): 661–63.

———. *The Gene Wars: Science, Politics, and the Human Genome*. New York: W. W. Norton, 1994.

Cooper, Melinda. *Life as Surplus: Biotechnology and Capitalism in the Neoliberal Era*. Seattle: University of Washington Press, 2008.

Cooper, Richard, Charles Rotimi, Susan Ataman, Daniel McGee, Babatunde Osotimehin, Solomon Kadiri, Walinjom Muna, et al. "The Prevalence of Hypertension in Seven Populations of West African Origin." *American Journal of Public Health* 87, no. 2 (1997): 160–68.

Couzin, Jennifer. "New Mapping Project Splits the Community." *Science* 296 (2002): 1391–93.

Creager, Angela. "Mobilizing Biomedicine: Virus Research between Lay Health Organizations and the U.S. Federal Government, 1935–1955." In *Biomedicine in the Twentieth Century: Practices, Policies, and Politics*, edited by Caroline Hannaway, 171–201. Amsterdam: IOS Press, 2008.

Cukier, Kenneth Neil. "Open Source Biotech: Can a Non-Proprietary Approach to Intellectual Property Work in the Life Sciences?" *Acumen Journal of Life Sciences* 1 (2003): 3.

Darling, Katherine Weatherford, Sara L. Ackerman, Robert H. Hiatt, Sandra Soo-Jin Lee, and Janet K. Shim. "Enacting the Molecular Imperative: How Gene-Environment Interaction Research Links Bodies and Environments in the Post-Genomic Age." *Social Science and Medicine* 155 (2016): 51–60. doi: http://dx.doi.org/10.1016/j.socscimed.2016.03.007.

Darnovsky, Marcy. "The Spitterati and Genome Research." *Spero News*, December 9, 2008. Accessed August 18, 2010. http://www.speroforum.com/a/17030/The-Spitterati-and-genome-research.

Daston, Lorraine. "Objectivity and the Escape from Perspective." In *The Science Studies Reader*, edited by Mario Biagioli, 110–23. London: Routledge, 1992.

Dausset, Jean, Howard Cann, Daniel Cohen, Mark Lathrop, Jean-Marc Lalouel, and Ray White. "Centre D'etude du Polymorphisme Humain (Ceph): Collaborative Genetic Mapping of the Human Genome." *Genomics* 6 (1990): 575–77.

Davie, George Elder. *The Democratic Intellect: Scotland and Her Universities in the Nineteenth Century*. Edinburgh: University of Edinburgh Press, 1961.

Davies, Kevin. "Keeping Score of Your Sequence." Accessed April 18, 2010. http://www.bio-itworld.com/BioIT_Article.aspx?id=84352.

Dawson, Cragg Ross. *Public Perceptions on the Collection of Human Biological Samples*. London: Wellcome Trust/MRC, 2000.

Dean, Jodi. *Democracy and Other Neoliberal Fantasies: Communicative Capitalism and Left Politics*. Durham, NC: Duke University Press, 2009.

———. *Publicity's Secret: How Technoculture Capitalizes on Democracy*. Ithaca, NY: Cornell University Press, 2002.

Dembner, Alice. "Harvard-Affiliated Gene Studies in China Face Federal Inquiry." *Boston Globe*, August 1, 2000. http://ahrp.org/harvard-affiliated-gene-studies-in-china-face-federal-inquiry/.

Desmond-Hellman, Susan. "Attention Stressed Out Doc: Can the Consumer Be the 'Cavalry' That Saves You?" Lecture, 2013 TedMed talk. http://www.tedmed.com/talks/show?id=18047.

———. "Toward Precision Medicine: A New Social Contract?" *Science Translational Medicine* 4, no. 129 (2012).

Diamond v. Chakrabarty, 447 U.S. 303 (1980).

Dickson, Samuel P., Kai Wang, Ian Krantz, Hakon Hakonarson, and David B. Goldstein. "Rare Variants Create Synthetic Genome-Wide Associations." *PLoS Biology* 8 (2010): 1.

Disch, Lisa. "More Truth Than Fact: Storytelling as Critical Understanding in the Writings of Hannah Arendt." *Politcal Philosophy* 21, no. 4 (1993): 665–94.

DIYBio. "An Institution for the Do-It-Yourself Biologist." Accessed February 2, 2015. https://diybio.org/.

DNAnexus. "DNAnexus Launches Web-Based Service to Ease Data Bottleneck in Next-Generation DNA Sequencing," *Business Wire*, accessed January 27, 2015, http://www.businesswire.com/news/home/20100420005195/en/DNAnexus-Launches-Web-Based-Service-Ease-Data-Bottleneck#.VMk07Ydijqk.

Donohue, "Beyond Ethics: The Scientific and Technological Development of the International HapMap Project 1998 to 2005." Paper presented to Capturing the History of Genomics: A Workshop, The National Human Genome Research Institute, April 29–30, 2015.

Du Bois, W. E. B. *The Souls of Black Folk*. New York: New American Library, 1903.

Duncan, Nigel. "World Medical Association Opposes Icelandic Gene Database." *British Medical Journal* 318, no. 7191 (1999): 1096. doi: 10.1136/bmj.318.7191.1096a.

Dunston, Georgia M. "G-RAP: A Model HBCU Genomic Research and Training Program." In *Plain Talk about the Human Genome Project: A Tuskegee University Conference on Its Promise and Perils and Matters of Race*, edited by Edward Smith and Walter Sapp, 51–58. Tuskegee, AL: Tuskegee University Press, 1997.

———. "A Passion for the Science of the Human Genome." *Molecular Biology of the Cell* 23, no. 21 (2012): 4154–56.

Durkheim, Émile. *Suicide: A Study in Sociology*. New York: Free Press, 1951.

Duster, Troy. *Backdoor to Eugenics*. London: Routledge, 1990.

———. "Review Essay in Symposium on the Bell Curve." *Contemporary Sociology* 24, no. 2 (1995): 158–61.

Eamon, William. *Science and the Secrets of Nature: Books of Secrets in Medieval and Early Modern Culture*. Princeton, NJ: Princeton University Press, 1994.

Ehley, Brianna. "Why Obama's $30B Digital Health Record Plan Is Failing." *Fiscal Times*, November 3, 2014. http://www.thefiscaltimes.com/2014/11/03/Why-Obama-s-30B-Digital-Health-Record-Plan-Failing.

Epstein, Ronald M. "Assessment in Medical Education." *New England Jounral of Medicine* 356 (2007): 387–96. doi: 10.1056/NEJMra054784.

Epstein, Steven. *Inclusion: The Politics of Difference in Medical Research*. Chicago: University of Chicago Press, 2007.

Etzioni, Amitai. "Justice: What's the Right Thing to Do? Review." *Hedgehog Review* 12, no. 1 (2010).

Eyfjörd, Jorunn Erla, Helga M. Ögmundsdottir, and Tomas Zoëga. "Decode Deferred." *Nature Biotechnology* 16 (1998): 491.

Ezrahi, Yaron. *The Descent of Icarus*. Cambridge, MA: Harvard University Press, 1990.

Fahey, Ciarán. "How Berliners Refused to Give Tempelhof Airport over to Developers." *The Guardian*, March 5, 2015. https://www.theguardian.com/cities/2015/mar/05/how-berliners-refused-to-give-tempelhof-airport-over-to-developers.

Fears, Robin, and George Poste. "Building Population Genetics Resources Using the U.K. NHS." *Science* 284, no. 5412 (1999): 267–68. doi: 10.1126/science.284.5412.267.

Fischer, Claude S., Michael Hout, Martin Sanchez Jankowski, Samuel R. Lucas, Ann Swidler, and Kim Voss. *Inequality by Design: Cracking the Bell Curve Myth*. Princeton, NJ: Princeton University Press, 1996.

Fitzpatrick, Tony. "Biological Differences among Races Do Not Exist, WU Research Shows." *Washington University Record*, October 15, 1998.

Fleck, Leonard, John Castillo, Frances Krouse, E. Yvonne Lewis, Shearon Smith, Daniel Soza, and Brian K. Sydnor. "Summary Dialogue Report." *Communities of Color and Genetics Policy Project* (2006). http://sph.umich.edu/genpolicy/current/reports/summary_dialogue_report.pdf, accessed October 21, 2015.

Fletcher, Amy Lynn. "Field of Genes: The Politics of Science and Identity in the Estonian Genome Project." *New Genetics and Society* 23, no. 1 (2004): 3–14.

Flint, Anthony. "Don't Classify by Race, Urge Scientists." *Boston Globe*, March 5, 1995.

Foner, Eric. *The Story of American Freedom*. New York: W. W. Norton, 1999.

Fortun, Michael A. "The Human Genome Project and the Acceleration of Biotechnology." In *Private Science: Biotechnology and the Rise of the Molecular Sciences*, edited by Arnold Thackray, 182–201. Philadelphia: University of Pennsylvania Press, 1998.

———. *Promising Genomics: Iceland and DeCODE Genetics in a World of Speculation*. Berkeley: University of California Press, 2008.

Foucault, Michel. "'Omnes et Singulatim': Toward a Critique of Political Reason." In *Power: Essential Works of Foucault, 1954–1984*, vol. 3, edited by James D. Faubion, 298–325. New York: New Press, 1997.

Fournier, Ron, and Sophie Quinton. "How Americans Lost Trust in Our Greatest Institutions." *The Atlantic*, April 12, 2012. http://www.theatlantic.com/politics/archive/2012/04/how-americans-lost-trust-in-our-greatest-institutions/256163/.

Franklin, Sarah. "Life Itself: Global Nature and the Genetic Imaginary." In *Global Nature, Global Culture*, edited by Sarah Franklin, Celia Lurie, and Jackie Stacey, 188–227. London: Sage, 2000.

Franklin, Seb. "Virality, Informatics, and Critique; or, Can There Be Such a Thing as Radical Computation?" *Women's Studies Quarterly* 40, no. 1/2 (2012): 153–70.

Freedland, Jonathan. "Post-Truth Politicians Such as Donald Trump and Boris Johnson Are No Joke." *The Guardian*, May 13, 2016. https://www.theguardian.com/commentisfree/2016/may/13/boris-johnson-donald-trump-post-truth-politician.

Friedlander, Jonathan. "Genes, People, and Property: Furor Erupts over Genetic Research on Indigenous Groups." *Cultural Survival* 20, no. 2 (1995): 22–24.

Galison, Peter. *How Experiments End*. Chicago: University of Chicago Press, 1987.

Generation Scotland. *Generation Scotland Management, Access, and Policy Publication*. Edinburgh: Generation Scotland, 2007. http://www.ed.ac.uk/files/atoms/files/gs_mapp.pdf.

Genome Biology Editorial Team. "Closure of the NCBI SRA and Implications for the Long-Term Future of Genomics and Data Storage." *Genome Biology* 12, no. 3 (2011): 402. http://www.ncbi.nlm.nih.gov/pmc/articles/PMC3129670/.

GenomeWeb. "NIH to Pump up to $96m into New Big Data Centers." *Genome Web Daily News*, July 23, 2013. https://www.genomeweb.com/informatics/nih-pump-96m-new-big-data-centers.

———. "Walgreens to Sell Pathway Genomics' Sample Collection Kit." *Genome Web*, May 11, 2010. Accessed August 2010.

"Genomics Gets Personal: Property, Persons, Privacy." YouTube Video. Posted by University of California–Santa Cruz, November 1, 2012. http://www.youtube.com/watch?v=56nIQ68wV84.

Gertner, Jon. *The Idea Factory: Bell Labs and the Great Age of American Innovation*. New York: Penguin, 2012.

Gilroy, Paul. *Against Race: Imagining Political Culture beyond the Color Line*. Cambridge, MA: Harvard University Press, 2000.

Gitschier, Jane. "Inferential Genotyping of Y Chromosomes in Latter-Day Saints Founders and Comparison to Utah Samples in the HapMap Project." *American Journal of Human Genetics* 84, no. 2 (2009): 251–58.

Gleick, James. *The Information: A History, a Theory, a Flood.* New York: Pantheon, 2011.

Goetz, Thomas. "Sergey Brin's Search for a Parkinson's Cure." *Wired*, June 22, 2010. Accessed October 13, 2014. http://www.wired.com/magazine/2010/06/ff_sergeys_search/.

Goldberg, David Theo. *The Threat of Race: Reflections on Racial Neoliberalsim.* Malden, MA: Blackwell, 2009.

Gordon, Avery. *Ghostly Matters: Haunting and the Sociological Imagination.* Minneapolis: University of Minnesota Press, 2008.

Gordon, Jessica Michelle. "Why Estonia? A Population's Self-Selection for National Genetic Research." Master's thesis, Harvard University, 2003.

Gottweis, Herbert, and Alan Peterson, eds. *Biobanks: Governance in Comparative Perspective.* London: Routledge, 2008.

Graber, Cynthia. "The Problem with Precision Medicine." *New Yorker*, February 5, 2015. http://www.newyorker.com/tech/elements/problem-precision-medicine.

Green, Richard E., Johannes Krause, Adrian W. Briggs, Tomislav Maricic, Udo Stenzel, Martin Kircher, Nick Patterson, et al. "A Draft Sequence of the Neandertal Genome." *Science* 328, no. 5979 (2010): 710–22. doi: 10.1126/science.1188021.

Green, Robert C., J. Scott Roberts, L. Adrienne Cupples, Norman R. Relkin, Peter J. Whitehouse, Tamsen Brown, Susan LaRusse Eckert, et al. "Disclosure of *APOE* Genotype for Risk of Alzheimer's Disease." *New England Journal of Medicine* 361, no. 3 (2009): 245–54. doi: 10.1056/NEJMoa0809578.

Greenberg, Judith. "Special Oversight Groups to Add Protections for Population-Based Repository Samples." *American Journal of Human Genetics* 66 (2000): 745–47.

Greene, David. "Seven Decades Ago, a New, Enormous Kind of Explosion." July 17, 2015. http://www.npr.org/2015/07/17/423740547/seven-decades-ago-a-new-enormous-kind-of-explosion.

Gross, Liza. "De-Extinction Debate: Should Extinct Species Be Received?" *KQED Science*, June 5, 2013. Accessed October 12, 2015. http://ww2.kqed.org/science/2013/06/05/deextinction-debate-should-extinct-species-be-revived/.

Guerrero Mothelet, Veronica, and Stephan Herrera. "Mexico Launches Bold Genome Project." *Nature Biotechnology* 23, no. 1030 (2005). doi: 10.1038/nbt0905-1030.

Gumz, Jondi. "UCSC Goes Bald to Back Childhood Cancer Research." Accessed October 14, 2016. https://genomics.soe.ucsc.edu/news/article/152.

Guthrie, Woody. *This Land Is Your Land: The Asch Recordings Volume 1*, track 14. Smithsonian Folkways Recordings, 1997.

Hacking, Ian. *The Social Construction of What?* Cambridge, MA: Harvard University Press, 1999.

Haddow, Gill, Lorraine Murray, Katherine Myant, and Anna Carlson. *Public Attitudes to Participation in Generation Scotland*. Edinburgh: INNOGEN/ Ipsos MORI, 2007.

Haddow, Gill, Sarah Cunningham-Burley, Ann Bruce, and Sarah Parry. "Generation Scotland: Consulting Publics and Specialists at an Early Stage in a Genetic Database's Development." *Critical Public Health* 18, no. 2 (2008): 139–49.

Haley, Alex. *Roots: The Saga of an American Family*. New York: Doubleday Books, 1976.

Hamblin, Martha, and Anna Di Rienzo. "Detection of the Signature of Natural Selection in Humans: Evidence from the Duffy Blood Group Locus." *American Journal of Human Genetics* 66, no. 5 (2000): 1669–79.

Haraway, Donna. "Anthropocene, Capitalocene, Plantationocene, Chthulucene: Making Kin." *Environmental Humanities* 6 (2015): 159–65.

———. "Awash in Urine: Des and Premarin® in Multispecies Response-Ability." *Women's Studies Quarterly* 40, no. 1 (2012): 301–16.

———. *Simians, Cyborgs, and Women*. New York: Routledge, 1991.

———. *When Species Meet*. Minneapolis: University of Minnesota Press, 2008.

Harris, Marilynn. "Letter to the Editor." *San Francisco Chronicle*, March 8, 2016. Accessed January 27, 2015. http://www.sfgate.com/opinion/letterstoeditor /article/Letters-to-the-editor-March-7-4339359.php.

Harrison, Charlotte. "Neither Moore nor the Market: Alternative Models for Compensation Contributors of Human Tissue." *American Journal of Law and Medicine* 28 (2002): 77–105.

Harry, Debra. "The Human Genome Diversity Project and Its Implications for Indigenous Peoples." Accessed May 13, 2013. http://www.ipcb.org/publica tions/briefing_papers/files/hgdp.html.

———. "Patenting Life and Its Implications for Indigenous Peoples." *Information about Intellectual Property Rights* 7 (1995): 1–2.

Hartocollis, Anemona. "Cancer Centers Racing to Map Patients' Genes." *New York Times*, April 21, 2013. http://www.nytimes.com/2013/04/22/health /patients-genes-seen-as-future-of-cancer-care.html.

Hayden, Cori. "Taking as Giving: Bioscience, Exchange, and the Politics of Benefit-Sharing." *Social Studies of Science* 37, no. 5 (2007): 729–58.

Hayden, Erika Check. "Genome Researchers Raise Alarm over Big Data." *Nature*, July 7, 2015. http://www.nature.com/news/genome-researchers-raise-alarm -over-big-data-1.17912.

———. "Genome Sequencing Stumbles towards the Clinic." *Nature*, March 11, 2014.

———. "Technology: The $1,000 Genome." *Nature*, March 19, 2014. http://www .nature.com/news/technology-the-1-000-genome-1.14901.

Hazell, Robert, ed. *The State and the Nations: The First Year of Devolution in the United Kingdom*. Thorverton, UK: Imprint Academic, 2000.

Herrnstein, Richard, and Charles Murray. *The Bell Curve: Intelligence and Class Structure in American Life*. New York: Free Press, 1994.

Hickman, Leo. "GM Crops: Protesters Go Back to the Battlefields." *The Guardian*, May 22, 2012. https://www.theguardian.com/environment/2012/may/22/gm-crops-protesters-battlefields.

Higginbotham, Evelyn Brooks. "African-American Women's History and the Metalanguage of Race." *Signs: A Journal of Women in Culture and Society* 17 (1992): 251–74.

Hilgartner, Stephen. "Selective Flows of Knowledge in Technoscientific Interaction: Information Control in Genome Research." *British Society for the History of Science* 45, no. 2 (2012): 267–80.

Horton, Carol A. *Race and the Making of American Liberalism*. Oxford: Oxford University Press, 2005.

Hotz, Robert Lee. "Scientists Say Race Has No Biological Basis." *Los Angeles Times*, February 20, 1995. http://articles.latimes.com/1995-02-20/news/mn-34098_1_biological-basis.

House of Lords. *Science and Society, Third Report of Session*. London: House of Lords Select Committee on Science and Technology, HL38, 2000.

Huet, Ellen. "Google Bus Blocked by Dancing Protesters in S.F. Mission District." *SF Gate*, April 1, 2014. Accessed April 1, 2014. http://blog.sfgate.com/techchron/2014/04/01/google-bus-blocked-by-dancing-protesters-in-s-f-mission-district/#22081101=0.

Human Genome News. "Ramping Up Production Sequencing." *Human Genome News* 8, no. 2 (1996): 3.

Humphreys, West H. *Reports of Cases Argued and Determined in the Supreme Court of Tennessee*. Vol. 4. Saint Louis, MO: Soule, Thomas, and Winsor, 1871.

Hunter, David J., David Altshuler, and Daniel J. Rader. "From Darwin's Finches to Canaries in the Coal Mine—Mining the Genome for New Biology." *New England Journal of Medicine* 358 (2008): 2760–63.

Hunter, David J., Muin J. Khoury, and Jeffrey M. Drazen. "Letting the Genome out of the Bottle—Will We Get Our Wish?" *New England Journal of Medicine* 358, no. 2 (2008): 105–7. doi: 10.1056/NEJMp0708162.

Indigenous Peoples Council on Biocolonialism. "Resolution by Indigenous Peoples." Accessed March 5, 2002. http://www.ipcb.org/resolutions/index.htm.

International HapMap Consortium. "Integrating Ethics and Science in the International HapMap Project." *Nature Reviews Genetics* 5 (2004): 467–75.

———. "The International HapMap Project." *Nature* 426 (2003): 789–93.

Jack, Ian. "Free Education for Young Scots? Only If the English Students Pay Full Whack." *The Guardian*, March 26, 2011.

Jackson, Fatimah. "Assessing the Human Genome Project: An African American and Bioanthropological Critique." In *Plain Talk about the Human Genome Project: A Tuskegee University Conference on Its Promise and Perils and Matters of Race*, edited by Edward Smith and Walter Sapp, 105–12. Tuskegee, AL: Tuskegee University Press, 1997.

Jaffe, Susan. "End in Sight for Revision of U.S. Medical Research Rules." *Lancet* 386, no. 10000 (2015): 1225–26. doi: 10.1016/S0140-6736(15)00314-1.

Jameson, Fredric. *The Political Unconscious: Narrative as a Socially Symbolic Act*. Ithaca, NY: Cornell University Press, 1982.

Jasanoff, Sheila. *Designs on Nature: Science and Democracy in Europe and the United States*. Princeton, NJ: Princeton University Press, 2005.

———. *The Ethics of Invention: Technology and the Human Future*. New York: Norton, 2016.

———. "The Idiom of Co-Production." In *States of Knowledge: The Co-Production of Science and Social Order*, edited by Shelia Jasanoff. London: Routledge, 2004.

———. *Science at the Bar: Law, Science, and Technology in America*. Cambridge, MA: Harvard University Press, 1995.

Jasanoff, Sheila, ed. *Reframing Rights: Bioconstitutionalism in the Genetic Age*. Cambridge, MA: MIT Press, 2011.

Jenkins, Henry. *Convergence Culture: Where Old Media and New Media Collide*. New York: New York University Press, 2006.

Johns, Adrian. *Piracy: The Intellectual Property Wars from Gutenberg to Gates*. Chicago: University of Chicago Press, 2009.

Johnson, Carolyn Y. "George Church, Harvard Genetics Professor, Banters with Stephen Colbert." *Boston Globe*, October 5, 2012. Accessed June 26, 2015. http://www.boston.com/whitecoatnotes/2012/10/05/george-church-harvard-genetics-professor-banters-with-stephen-colbert/IqXfHArTmkxfJPqM1rfM0O/story.html.

Jones, James H. *Bad Blood: The Tuskegee Syphilis Experiment*. New York: Free Press, 1981.

Jones, Kathryn Maxson, Rachel A. Ankeny, and Robert Cook-Deegan. "The Bermuda Triangle: The Politics, Principles, and Pragmatics of Data Sharing in the History of the Human Genome Project." *Journal of the History of Biology* (forthcoming).

Jones, Orion. "How Social Media Created an Echo Chamber for Ideas." Accessed October 21, 2015. http://bigthink.com/ideafeed/how-social-media-have-created-echo-chambers-for-ideas.

Jones, Reece. "Death in the Sands: The Horror of the U.S. Mexico Border." *The Guardian*, October 4, 2016. https://www.theguardian.com/us-news/2016/oct/04/us-mexico-border-patrol-trump-beautiful-wall.

Joyner, Michael J., Nigel Paneth, and John A. Ioannidis. "What Happens When Underperforming Big Ideas in Research Become Entrenched?" *JAMA* (October 4, 2016). doi: 10.1001/jama.2016.11076.

Kahn, Patricia. "Genetic Diversity Project Tries Again." *Science* 266 (1994): 720–22.

Kanter, James, and Somini Sengupta. "Europe Continues Wrestling with Online Privacy Rules." *New York Times*, June 6, 2013. Accessed June 7, 2013. http://www.nytimes.com/2013/06/07/technology/europe-still-wrangling-over-online-privacy-rules.html?pagewanted=1&_r=0&hp.

Kay, Lily. *Who Wrote the Book of Life? A History of the Genetic Code*. Palo Alto, CA: Stanford University Press, 2000.

Kaye, Jane, and Mark Stranger, eds. *Principles and Practice in Biobank Governance*. London: Ashgate, 2009.

Keating, Michael. *The Government of Scotland: Public Policy Making after Devolution*. Edinburgh: Edinburgh University Press, 2005.

Kelleher, Kevin. "Google: World Domination Starts Today." *Pando*, October 18, 2013. Accessed October 11, 2014. http://pando.com/2013/10/18/google-world-domination-starts-today/.

Keller, Evelyn Fox. *Secrets of Life, Secrets of Death: Essays on Language, Gender, and Science*. New York: Routledge, 1992.

Kelty, Christopher M. "This Is Not an Article: Model Organism Newsletters and the Question of 'Open' Science." *Biosocieties* 7, no. 2 (2012): 140–68.

———. *Two Bits: The Cultural Significance of Free Software*. Durham, NC: Duke University Press, 2008.

Kennedy, Donald. "Not Wicked, Perhaps, but Tacky." *Science* 297, no. 5585 (2002): 1237.

Kevles, Daniel. *In the Name of Eugenics: Genetics and the Uses of Human Heredity*. Berkeley: University of California Press, 1985.

Kim, Leland. "UCSF Wins Four CASE Awards of Excellence." *UCSF*, February 19, 2014. Accessed May 6, 2016. https://www.ucsf.edu/news/2014/02/124756/ucsf-wins-four-case-awards-excellence.

King, Patricia. "The Dilemma of Difference." In *Plain Talk about the Human Genome Project: A Tuskegee University Conference on Its Promise and Perils and Matters of Race*, edited by Edward Smith and Walter Sapp, 75–82. Tuskegee, AL: Tuskegee University Press, 1997.

———. *Interview with Patricia King: Belmont Oral History Project*. By LeRoy B. Walters. Georgetown University, September 9, 2004.

Kitcher, Phillip. *Science, Truth, and Democracy*. Oxford: Oxford University Press, 2001.

Klein, Naomi. *The Shock Doctrine: The Rise of Disaster Capitalism*. New York: Picador, 2008.

Kneebone, Elizabeth, Carey Nadeau, and Alan Berube. "The Re-Emergence of Concentrated Poverty: Metropolitan Trends in the 2000s." Metropolitan Policy Programs, Brookings, 2011.

Knight, Ben. "Berlin to Build on Tempelhof Despite Drop in Refugees." *DW*, March 25, 2016. Accessed August 24, 2016. http://www.dw.com/en/berlin-to-build-on-tempelhof-despite-drop-in-refugees/a-19143212.

Knight, Heather. "In Growth of Wealth Gap, We're No. 1." *SF Gate*, March 2, 2014. Accessed April 1, 2014. http://www.sfgate.com/bayarea/article/In-growth-of-wealth-gap-we-re-No-1-5281174.php.

Knight Foundation. "Open Humans Network." Accessed February 2, 2015. http://www.knightfoundation.org/grants/201447980/.

Koenig, Barbara A. "Have We Asked Too Much of Consent?" *Hastings Center Report* 44, no. 4 (2014): 33–34. doi: 10.1002/hast.329.

Kohler, Robert. *Lords of the Fly*: Drosophila *Genetics and the Experimental Life*. Chicago: University of Chicago Press, 1994.

Kokalitcheva, Kia. "Is Facebook a Political Echo Chamber? Social Network Says No." Accessed October 21, 2015. http://fortune.com/2015/05/07/facebook-echo-chamber/.

Kroll, David. "Obama's Precision Medicine Initiative: Paying for Precision Drugs Is the Challenge." *Forbes*, January 20, 2015. http://www.forbes.com/forbes/welcome/?toURL=http://www.forbes.com/sites/davidkroll/2015/01/20/obamas-admirable-precision-medicine-initiative-paying-for-precision-drugs-is-the-real-challenge/&refURL=https://www.google.com/&referrer=https://www.google.com/.

Kruglyak, Leonid. "Prospects for Whole-Genome Linkage Disequilibrium Mapping of Common Disease Genes." *Nature Genetics* 22 (1999): 139–44.

Kühl, Stefan. *The Nazi Connection: Eugenics, American Racism, and German National Socialism*. Oxford: Oxford University Press, 1994.

Lambro, Donald. *Fat City: How Washington Wastes Your Taxes*. South Bend, IN: Regenry/Gateway, 1980.

Landecker, Hannah. "Between Beneficence and Chattel: The Human Biological in Law and Science." *Science in Context* 12 (1999): 203–25.

Latour, Bruno. *Science in Action: How to Follow Scientists and Engineers through Society*. Cambridge, MA: Harvard University Press, 1987.

Latour, Bruno, and Peter Weibel, eds. *Making Things Public: Atmospheres of Democracy*. Cambridge, MA: MIT Press, 2005.

Laurie, Graeme T. *Genetic Privacy: A Challenge to Medico-Legal Norms*. Cambridge: Cambridge University Press, 2002.

Laurie, Graeme T., and Johanna Gibson. *Generation Scotland: Legal and Ethical Aspects*. Edinburgh: Arts and Humanities Research Council Research Centre for Studies in Intellectual Property and Technology Law, 2003.

Levy, Samuel, Granger Sutton, Pauline C. Ng, Lars Feuk, Aaron L. Halpern, Brian P. Walenz, Nelson Axelrod, et al. "The Dilpoid Genome Sequence of an Indivdual Human." *PLOS Biology* (2007). Accessed September 12, 2014. doi: www.plosbiology.org/article/info%3Adoi%2F10.1371%2Fjournal.pbio.0050254#authcontrib.

Lewin, Roger. "Genes from a Disappearing World." *New Scientist* (1993): 25–29.

Lewontin, Richard. "The Apportionment of Human Diversity." *Evolutionary Biology* 6 (1972): 381–98.

———. "People Are Not Commodities." *New York Times*, January 23, 1999. https://www.ncbi.nlm.nih.gov/pubmed/11647693.

The Life and Times of Lilly Mendel. http://lillymendel.blogspot.co.uk/2009/11/inspiration-comes-in-many-forms-idea-to.html.

Lincoln, Abraham. "The Gettysburg Address." Dedication of the Soldiers' National Cemetery, Gettysburg. November 19, 1863.

Lindee, M. Susan. "Survivors and Scientists: Hiroshima, Fukushima, and the Radiation Effects Research Foundation, 1975–2014." *Social Studies of Science* 46, no. 2 (2016): 184–209.

Lipset, Seymour Martin, and William Schneider. "The Decline of Confidence in American Institutions." *Political Science Quarterly* 98, no. 3 (1983): 379–402.

Liptak, Adam. "Justices Allow DNA Collection after an Arrest." *New York Times*, June 3, 2013. http://www.nytimes.com/2013/06/04/us/supreme-court-says-police-can-take-dna-samples.html.

Locke, John. *Two Treatises of Government*. Edited by R. Cox. Arlington Heights, IL: Harlan Davidson, 1982.

Lorenzen, Eline D., David Nogues-Bravo, Ludovic Orlando, Jaco Weinstock, Jonas Binladen, Katharine A. Marske, Andrew Ugan, et al. "Species-Specific Responses of Late Quaternary Megafauna to Climate and Humans." *Nature* 479, no. 7373 (2011): 359–64. doi: http://www.nature.com/nature/journal/v479/n7373/abs/nature10574.html—supplementary-information.

Lowe, Felix. "Google Wife Targets World DNA Domination." *Telegraph*, January 25, 2008. http://www.telegraph.co.uk/finance/newsbysector/mediatechnologyandtelecoms/2783261/Davos-2008-Google-wife-targets-world-DNA-domination.html.

Ludka, Alexandria. "Meet the New Facebook Millionares." *ABC News*, May 16, 2012. Accessed October 23, 2015. http://abcnews.go.com/Technology/facebook-millionaires/story?id=15499090.

Lunshof, Jeantine E., Jason Bobe, John Aach, Misha Angrist, Joseph V. Thakuria, Daniel B. Vorhaus, Margret R. Hoehe, et al. "Personal Genomes in Progress: From the Human Genome Project to the Personal Genome Project." *Dialogues in Clinical Neuroscience* 12, no. 1 (2010): 47–60.

Lunshof, Jeantine E., Ruth Chadwick, Daniel B. Vorhaus, and George M. Church. "From Genetic Privacy to Open Consent." *Nature Reviews Genetics* 9 (2008): 406–11.

Lyotard, Jean-François. *The Postmodern Condition: A Report on Knowledge*. Minneapolis: University of Minnesota Press, 1979.
Macer, Darryl. "Ethical Considerations in the HapMap Project." *Eubios Journal of Asian & International Bioethics* 13 (2003): 125–17.
Macilwain, Colin. "Diversity Project 'Does Not Merit Federal Funding.'" *Nature* 389 (1997): 774.
MacKenzie, Adrian. "Machine Learning and Genomic Dimensionality." In *Postgenomics: Perspectives on Biology*, edited by Sarah S. Richardson et al., 73–102. Durham, NC: Duke University Press, 2015.
Manrai, Arjun K., Birgit H. Funke, Heidi L. Rehm, Morten S. Olesen, Bradley A. Maron, Peter Szolovits, David M. Margulies, et al. "Genetic Misdiagnoses and the Potential for Health Disparities." *New England Journal of Medicine* 375, no. 7 (2016): 655–65. doi: 10.1056/NEJMsa1507092.
Marks, Jonathan. "Letter: The Trouble with the HGDP." *Molecular Medicine Today* (1998): 243.
Marshall, Eliot. "Tapping Iceland's DNA." *Science* 278, no. 5338 (1997): 566. doi: 10.1126/science.278.5338.566.
McCormick, Danny, David H. Bor, Stephanie Woolhandler, and David U. Himmelstein. "Giving Office-Based Physicians Electronic Access to Patients' Prior Imaging and Lab Results Did Not Deter Ordering of Tests." *Health Affairs* 31, no. 3 (2012): 488–96. doi: 10.1377/hlthaff.2011.0876.
McElheny, Victor K. *Drawing the Map of Life: Inside the Human Genome Project*. New York: Basic Books, 2010.
McKie, Robin. "Steam Rises over Iceland Gene Pool." *Observer*, November 8, 1998.
Meek, James. "Decode Was Meant to Save Lives . . . Now It's Destroying Them." *The Guardian*, October 31, 2002. https://www.theguardian.com/science/2002/oct/31/genetics.businessofresearch.
Merton, Robert. "Notes on Science and Democracy." *Journal of Legal and Political Sociology* 1, no. 1/2 (1942): 115–26.
Meyer, Michelle N. "Ethical Concerns, Conduct, and Public Policy for Re-Identification and De-Identification Practices." *Bill of Health Blog*, October 2, 2013. http://blogs.law.harvard.edu/billofhealth/category/re-identification-symposium/.
———. "Reflections of a Re-Identification Target, Part 1." *Bill of Health Blog*, May 24, 2013. http://blogs.law.harvard.edu/billofhealth/2013/05/24/reflections-of-a-re-identification-target-part-i-some-information-doesnt-want-to-be-free-re-identification-symposium/#more-6244.
Moore, Kelly. *Disrupting Science: Social Movements, American Scientists, and the Politics of the Military, 1945–1975*. Princeton, NJ: Princeton University Press, 2008.
Muller-Hill, Benno. *Murderous Science: Elimination by Scientific Selection of Jews, Gypsies, and Others, Germany 1933–1945*. Oxford: Oxford University Press, 1998.

Munro, Dan. "Class Action Law Suit Filed against 23andme." *Forbes*, December 2, 2013. http://www.forbes.com/forbes/welcome/?toURL=http://www.forbes.com/sites/danmunro/2013/12/02/class-action-law-suit-filed-against-23andme/&refURL=https://www.google.com/&referrer=https://www.google.com/.

Nelkin, Dorothy, and M. Susan Lindee. *The DNA Mystique: The Gene as a Cultural Icon*. Ann Arbor: University of Michigan Press, 2004.

Nelson, Alondra. *The Social Life of DNA: Race, Reparations, and Reconciliation after the Genome*. Boston: Beacon Press, 2016.

New York Times Editorial Board. "Silicon Valley's Diversity Problem." *New York Times*, Ocotber 4, 2014. http://www.nytimes.com/2014/10/05/opinion/sunday/silicon-valleys-diversity-problem.html?_r=0.

Ng, Pauline C., Sarah S. Murray, Samuel Levy and J. Craig Venter. "An Agenda for Personalized Medicine." *Nature* 461 (2009): 724–26.

Nixon, Rob. *Slow Violence and Environmentalism of the Poor*. Cambridge, MA: Harvard University Press, 2011.

Nunberg, Geoffrey. "Data Deluge." *New York Times*, March 20, 2011. http://query.nytimes.com/gst/fullpage.html?res=9E06E5DA1F3EF933A15750C0A9679D8B63&pagewanted=all.

Nussbaum, Emily. "Say Everything: The Future Belongs to the Uninhibited." *New York Magazine*, February 12, 2007. Accessed January 19, 2015. http://www.nymag.com/news/features/27341.

Obama, Barack. "Remarks of the President in Welcoming Senior Staff and Cabinet Secretaries to the White House." Speech, Washington, DC, January 21, 2009. https://www.whitehouse.gov/the-press-office/remarks-president-welcoming-senior-staff-and-cabinet-secretaries-white-house.

O'Riordan, Kate. *The Genome Incorporated*. Surrey, UK: Ashgate, 2010.

Olson, Maynard. "Genome Centers: What Is Their Role?" In *Plain Talk about the Human Genome Project: A Tuskegee University Conference on Its Promise and Perils and Matters of Race*, edited by Edward Smith and Walter Sapp, 17–26. Tuskegee, AL: Tuskegee University Press, 1997.

Owens, Kelly, and Mary-Claire King. "Genomic Views of Human History." *Science* 286, no. 5439 (1999): 451–53.

Panofsky, Aaron L. *Misbehaving Science: Controversy and the Development of Behavior Genetics*. Chicago: University of Chicago, 2014.

"Patents, Indigenous Peoples, and Human Genetic Diversity." *ETC Group*. Last modified May 30, 1993. http://www.etcgroup.org/content/patents-indigenous-peoples-and-human-genetic-diversity.

Patil, DJ, and Stephanie Devaney, "Next Steps in Developing the Precision Medicine Initiative." The White House, https://www.whitehouse.gov/blog/2015/08/21/next-steps-developing-precision-medicine-initiative, August 21, 2015.

Paul, Diane. *Controlling Human Heredity: 1865 to the Present*. Atlantic Highlands, NJ: Humanities Press, 1995.

Pemberton, Stephen. *The Bleeding Disease: Hemophilia and the Unintended Consequences of Medical Progress*. Baltimore: Johns Hopkins University Press, 2011.

Perlroth, Nicole. "Ashley Madison Chief Steps Down after Data Breach." *New York Times*, August 28, 2015. Accessed September 21, 2015. http://www.nytimes.com/2015/08/29/technology/ashley-madison-ceo-steps-down-after-data-hack.html?_r=0.

Personalized Medicine Coalition. "Personal Genomics and Industry Standards: Scientific Validity." 2008. Accessed November 6, 2016.http://www.personalizedmedicinecoalition.org/Userfiles/PMC-Corporate/file/pmc_scientific_validity.pdf.

Peters, Justin. *The Idealist: Aaron Swartz and the Rise of Free Culture of the Internet*. New York: Scribner, 2013.

Petryna, Adriana. *Life Exposed: Biological Citizens after Chernobyl*. Princeton, NJ: Princeton University Press, 2002.

Philippidis, Alex. "Top 10 Biopharma Clusters." *GEN: Genetic Engineering and Biotechnology News*, March 10, 2014. http://www.genengnews.com/insight-and-intelligence/top-10-u-s-biopharma-clusters/77900061.

Picciano, Bob. "Why Big Data Is the New Natural Resource." *Forbes*, June 30, 2014. Accessed November 29, 2015. http://www.forbes.com/sites/ibm/2014/06/30/why-big-data-is-the-new-natural-resource/#1322d7566a84.

Polanyi, Michael. "The Republic of Science: Its Political and Economic Theory." *Minerva* 1 (1962): 54–72.

Pollack, Andrew. "Awaiting the Genome Payoff." *New York Times*, June 14, 2010. http://www.nytimes.com/2010/06/15/business/15genome.html?pagewanted=all.

———. "California Licenses 2 Companies to Offer Gene Services." *New York Times*, August 19, 2008. Accessed September 10, 2014. http://www.nytimes.com/2008/08/20/business/20gene.html?_r=0.

———. "F.D.A. Faults Companies on Unapproved Genetic Tests." *New York Times*, June 10, 2010. http://www.nytimes.com/2010/06/12/health/12genome.html.

———. "Google Co-Founder Backs Vast Parkinson's Study." *New York Times*, March 11, 2009. Accessed September 10, 2014. http://www.nytimes.com/2009/03/12/business/12gene.html?_r=1&sq=23andMe&st=cse&adxnnl=1&scp=12&adxnnlx=1247007643-XOQziP4oBzZGqpkemkv8VQ.

———. "The Wide, Wild World of Genetic Testing." *New York Times*, September 12, 2006. http://www.nytimes.com/2006/09/12/business/smallbusiness/12genetic.html.

Poster, Mark. *The Mode of Information: Poststructuralism and Social Context*. Chicago: University of Chicago Press, 1990.

Prainsack, Barbara, and Alena Buyx. *Solidarity and Biomedicine and Beyond*. Cambridge: Cambridge University Press, 2017.

"The Precision Medicine Initiative Cohort Program—Building a Research Foundation for 21st Century Medicine." Precision Medicine Initiative (PMI)

Working Group Report to the Advisory Committee to the Director, NIH (2015).

Puig de la Bellacasa, Maria. "Matters of Care in Technoscience: Assembling Neglected Things." *Social Studies of Science* 41, no. 1 (2012): 85–106. http://www.genomeweb.com/dxpgx/walgreens-sell-pathway-genomics-sample-collection-kit.

Putnam, Robert D. "Bowling Alone: America's Declining Social Capital." *Journal of Democracy* 6, no. 1 (1995.): 65–78.

Rabinow, Paul. "Artificiality and Enlightenment: From Sociobiology to Biosociality." In *The Science Studies Reader*, edited by Mario Biagioli, 407–16. New York: Routledge, 1999.

———. *French DNA: Trouble in Purgatory*. Chicago: University of Chicago Press, 1999.

Ramos, Erin M., Corina Din-Lovinescu, Jonathan S. Berg, Lisa D. Brooks, Audrey Duncanson, Michael Dunn, Peter Good, et al. "Characterizing Genetic Variants for Clinical Action." *American Journal of Medical Genetics, Part C, Seminars in Medical Genetics* 0, no. 1 (2014): 93–104. doi: 10.1002/ajmg.c.31386.

Ramsden, Edmund. "A Differential Paradox: The Controversy Surrounding the Scottish Mental Surveys of Intelligence and Family Size." *Journal of the History of the Behavioral Sciences* 43, no. 2 (2007): 109–34.

Rawls, John. *A Theory of Justice*. Cambridge, MA: Harvard University Press, 1999.

Reardon, Jenny. "The Democratic, Anti-Racist Genome? Technoscience at the Limits of Liberalism." *Science as Culture* 21, no. 1 (2012): 25–47.

———. "Democratic Mis-Haps: The Problem of Democratization in a Time of Biopolitics." *BioSocieties* 2 (2007): 239–56.

———. "Genomics' Problem of Communication." *Journal of Science Communication* 10, no. 3 (2011): 1–6.

———. "The Human Genome Diversity Project: A Case Study in Coproduction." *Social Studies of Science* 31, no. 3 (2001): 357–88.

———. "The 'Persons' and 'Genomics' of Personal Genomics." *Personalized Medicine* 8, no. 1 (2010): 95–107.

———. *Race to the Finish: Identity and Governance in an Age of Genomics*. Edited by Paul Rabinow. Princeton, NJ: Princeton University Press, 2005.

———. "Should Patients Understand That They Are Research Subjects?" *San Francisco Chronicle*, March 3, 2013. Accessed January 5, 2015. http://www.sfgate.com/opinion/article/Should-patients-understand-that-they-are-research-4321242.php.

Reardon, Jenny, Jacob Metcalf, Martha Kenney, and Karen Barad. "Science and Justice: The Trouble and the Promise." *Catalyst* 1, no. 1 (2015).

Reddy, Deepa. "Good Gifts for the Common Good: Blood and Bioethics in the Market of Genetic Research." *Cultural Anthropology* 22, no. 3 (2007): 429–72.

Regaldo, Antonio. “Apple Has Plans for Your DNA.” *MIT Technology Review*, May 5, 2016. https://www.technologyreview.com/s/537081/apple-has-plans-for-your-dna/.

———. “Geneticists Begin Tests of an Internet for DNA.” *MIT Technology Review*, 2014.

———. “Illumina Says 228,000 Human Genomes Will Be Sequenced This Year.” *MIT Technology Review*, September 24, 2014. https://www.technologyreview.com/s/531091/emtech-illumina-says-228000-human-genomes-will-be-sequenced-this-year/.

Reich, David E., and Eric S. Lander. “On the Allelic Spectrum of Human Disease.” *Trends in Genetics* 69 (2001): 124–37.

Reich, David, Michael A. Nalls, W. H. Linda Kao, Ermeg L. Akylbekova, Arti Tandon, Nick Patterson, James Mullikin, et al. “Reduced Neutrophil Count in People of African Descent Is Due to a Regulatory Variant in the Duffy Antigen Receptor for Chemokines Gene.” *PLOS Genetics* 5, no. 1 (2009).

Reverby, Susan. *Examining Tuskegee: The Infamous Syphilis Study and Its Legacy*. Chapel Hill, NC: University of North Carolina Press, 2009.

Richardson, Sarah S., and Hallam Stevens, eds. *Postgenomics: Perspectives on Biology after the Genome*. Durham, NC: Duke University Press, 2015.

Richtel, Matt. “The Battle of Mission Bay.” *New York Times*, September 6, 2015. http://www.nytimes.com/2015/09/06/business/a-basketball-arena-battles-for-san-franciscos-heart.html.

Rivero, Daniel. “Kuwait’s New DNA Collection Law Is Scarier Than We Ever Imagined.” *Fusion*, August 24, 2016. Accessed August 24, 2016. http://fusion.net/story/339574/kuwaits-new-dna-collection-law-is-scarier-than-we-ever-imagined/.

Rose, Hilary, and Steven Rose. *Genes, Cells, and Brains: The Promethean Promise of the New Biology*. New York: Verso, 2012.

Rose, Nikolas. *The Politics of Life Itself: Biomedicine, Power, and Subjectivity in the Twenty-First Century*. Princeton, NJ: Princeton University Press, 2007.

Rose, Nikolas, and Carlos Novas. “Biological Citizenship.” In *Global Assemblages: Technologies, Politics and Ethics as Anthropological Problems*, edited by Aihwa Ong and Stephen J. Collier, 439–63. Oxford: Blackwell, 2005.

Rousseau, Jean-Jacques. *Discourses on Political Economy and the Social Contract*. Oxford: Oxford University Press, 1994.

———. *The Social Contract*. Oxford: Oxford University Press, 1994.

Roy, Arundhati. “Can We Leave the Bauxite in the Mountain? Field Notes on Democracy.” Presentation at Science and Democracy Lecture and Panel Discussion, Harvard University, April 1, 2010. Accessed October 14, 2015. http://sts.hks.harvard.edu/events/lectures/roy-mountain.html.

Royal Society. *Science in Society: The Impact of the Five Year Kohn Foundation Funded Programme*. London: Royal Society, 2006.

Said, Carolyn. "Pfizer to Open Research Center in SF's Mission Bay." *San Francisco Chronicle*, July 22, 2011. http://www.sfgate.com/business/article/Pfizer-to-open-research-center-in-SF-s-Mission-Bay-2353818.php.

Salkin, Allen. "When in Doubt, Spit It Out." *New York Times*, September 12, 2008. http://www.nytimes.com/2008/09/14/fashion/14spit.html.

Sandel, Michael. *Justice: What's the Right Thing to Do?* New York: Farrar, Straus, and Giroux, 2009.

Sankaran, Neeraja. "African American Genome Mappers Pledge to Carry on Despite Grant Rejection." *Scientist* 9, no. 5 (1995): 1.

Scherer, Michael. "The Secret Sharers: A Close Look at Privacy, Surveillance, and Hacktivism." *Time*, June 24, 2013. http://content.time.com/time/magazine/article/0,9171,2145982,00.html.

Schultze, Charles L., William T. Dickens, and Thomas J. Kane. "Does the Bell Curve Ring True? A Closer Look at a Grim Portrait of American Society." *Brookings*, June 1, 1995. Accessed June 10, 2013. http://www.brookings.edu/research/articles/1995/06/does-the-bell-curve-ring-true.

Science and Justice Research Center (Collaborations) Group. "Experiments in Collaboration: Interdisciplinary Graduate Education in Science and Justice." *PLoS Biology* 11, no. 7 (2013): 1–5.

Sclove, Richard E. *Democracy and Technology*. New York: Guilford Press, 1995.

Scott, Joan. *Only Paradoxes to Offer: French Feminists and the Rights of Man*. Cambridge, MA: Harvard University Press, 1996.

Shapin, Steven. *The Scientific Life: A Moral History of a Late Modern Vocation*. Chicago: University of Chicago Press, 2008.

Shapin, Steven, and Simon Schaffer. *Leviathan and the Air-Pump: Hobbes, Boyle, and the Experimental Life*. Princeton, NJ: Princeton University Press, 1985.

Shapiro, B., and M. Hofreiter. "A Paleogenomic Perspective on Evolution and Gene Function: New Insights from Ancient DNA." *Science* 343, no. 6169 (2014). doi: 10.1126/science.1236573.

Shearlaw, Maeve. "Egypt Five Years On: Was It Ever a 'Social Media Revolution'?" *The Guardian*, January 25, 2016. Accessed October 16, 2016. https://www.theguardian.com/world/2016/jan/25/egypt-5-years-on-was-it-ever-a-social-media-revolution.

Shorett, Peter, Paul Rabinow, and Paul R. Billings. "The Changing Norms of the Life Sciences." *Nature Biotechnology* 21, no. 2 (2003): 123–25.

Shreeve, James. *The Genome War: How Craig Venter Tried to Capture the Code of Life and Save the World*. New York: Ballantine Books, 2005.

Shuren, Jeff. "Statement before the Subcommittee on Oversight and Investigations, Committee on Energy and Commerce, United States House of Representatives." Hearing transcript presented at Direct to Consumer Genetic Testing and the Consequences to the Public Health, Washington, DC, July 22, 2010.

Sigurdsson, Skúli. "Springtime in Iceland and 200 Million Dollar Promises." *Tageszeitung*, April 20, 2002.

Silverman, Rachel. "The Blood Group 'Fad' in Post-War Racial Anthropology." *Kroeber Anthropological Society Papers*. Edited by Jonathan Marks. Berkeley: University of California, 2000.

Skloot, Rebecca. *The Immortal Life of Henrietta Lacks*. New York: Crown Publishers, 2010.

———. "The Immortal Life of Henrietta Lacks, the Sequel." *New York Times*, May 23, 2013. http://www.nytimes.com/2013/03/24/opinion/sunday/the-immortal-life-of-henrietta-lacks-the-sequel.html.

Skrentny, John. *The Minority Rights Revolution*. Cambridge, MA: Harvard University Press, 2002.

Smith, Blair, Harry Campbell, Douglas Blackwood, John Connell, Mike Connor, Ian Deary, Anna Dominiczak, et al. "Generation Scotland: The Scottish Family Health Study; a New Resource for Researching Genes and Heritability." *BMC Medical Genetics* 7, no. 1 (2006): 74.

Smith, Edward. "Trickle-Down Genomics: Reforming 'Small Science' as We Know It." *Scientist* 13, no. 15 (1999): 19.

Sorensen, Eric. "Race Gene Does Not Exist, Say Scientists." *Seattle Times*, February 11, 2001. http://community.seattletimes.nwsource.com/archive/?date=20010211&slug=race11m.

Specter, Michael. "Decoding Iceland." *New Yorker*, January 18, 1999.

Spiegel. "Spiegel Interview with Craig Venter: 'We Have Learned Nothing from the Genome.'" *Spiegel*, July 29, 2010.

Starr, Paul. *Freedom's Power: The True Force of Liberalism*. New York: Basic Books, 2007.

Stepan, Nancy. *The Hour of Eugenics: Race, Gender, and Nation in Latin America*. Ithaca, NY: Cornell University Press, 1991.

Stevens, Hallam. *Life out of Sequence: A Data-Driven History of Bioinformatics*. Chicago: University of Chicago Press, 2013.

Stiglitz, Joseph E. *The Price of Inequality: How Today's Divided Society Endangers Our Future*. New York: W. W. Norton, 2012.

Strasser, Bruno J. "The Experimenter's Museum: Genbank, Natural History, and the Moral Economies of Biomedicine, 1979–1982." *Isis* 102 (2011): 60–96.

Sulston, John, and Georgina Ferry. *The Common Thread: A Story of Science, Politics, Ethics, and the Human Genome*. Washington, DC: John Henry Press, 2002.

Sunder Rajan, Kaushik. *Biocapital: The Constitution of Postgenomic Life*. Durham, NC: Duke University Press, 2006.

Swaine, Jon, Oliver Laughland, Jamiles Lartey, Ciara McCarthy. "Young Black Men Killed by US Police at Highest Rate in Year of 1,134 Deaths." *The Guardian*, December 13, 2015. https://www.theguardian.com/us-news/2015/dec/31/the-counted-police-killings-2015-young-black-men.

Swedlund, Alan. "Is There an Echo in Here? Historical Reflections on the Human Genome Diversity Project." Paper presented at the Wenner Gren Conference: Anthropological Perspectives on the Human Genome Diversity Project, Sever Springs Center, Mount Kisko, New York, November 3–7, 1993.

TallBear, Kimberly. "Narratives of Race and Indigeneity in the Genographic Project." *Journal of Law, Medicine, and Ethics* 35, no. 3 (2007): 412–24.

———. *Native American DNA: Tribal Belonging and the False Promise of Genetic Science*. Minneapolis: University of Minnesota Press, 2013.

Tanner, Adam. "Harvard Professor Re-Identifies Anonymous Volunteers in DNA Study." *Forbes*, April 25, 2013. http://www.forbes.com/sites/adamtanner/2013/04/25/harvard-professor-re-identifies-anonymous-volunteers-in-dna-study/#6fd68553e39f.

Tayag, Joe. "Toward Fair Cures: Diversity Policies in Stem Cell Research." *Greenlining Institute* (2006). Accessed October 16, 2015. doi: http://greenlining.org/wp-content/uploads/2013/02/TowardFairCuresDiversityPoliciesinStemCellResearch.pdf.

Teilhard de Chardin, Pierre. *The Future of Man*. New York: Doubleday, 1962.

Terranova, Tiziana. *Network Culture: Politics of the Information Age*. London: Pluto Press, 2004.

Thielman, Sam. "Libraries Promise to Destroy User Data to Avoid Threat of Government Surveillance." *The Guardian*, November 30, 2016. https://www.theguardian.com/books/2016/nov/30/library-user-data-government-surveillance-donald-trump.

Thies, Jochen. *Hitler's Plans for Global Domination: Nazi Architecture and Ultimate War Aims*. Oxford: Berghahn Books, 2012.

Thompson, Charis. *Good Science: The Ethical Choreography of Stem Cell Research*. Cambridge, MA: MIT Press, 2013.

Thompson, E. P. *The Making of the the English Working Class*. London: Victor Gollancz, 1963.

Thompson, Edward P. "The Moral Economy Reviewed." In *Customs in Common*. London: Merlin Press, 1991.

Thurston, Robin. "Why I'm Joining Helix as CEO." *Linked In*, July 12, 2016. https://www.linkedin.com/pulse/why-im-joining-helix-ceo-robin-thurston.

Timmerman, Luke. "DNA Sequencing Market Will Exceed $20 Billion, Says Illumina CEO Jay Flatley." *Forbes*, April 29, 2015. http://www.forbes.com/forbes/welcome/?toURL=http://www.forbes.com/sites/luketimmerman/2015/04/29/qa-with-jay-flatley-ceo-of-illumina-the-genomics-company-pursuing-a-20b-market/&refURL=https://www.google.com/&referrer=https://www.google.com/.

———. "Stephen Friend, Leaving High Powered Merck Gig, Lights Fire for Open Source Biology Movement." *Xconomy*, August 6, 2009. Accessed February 2, 2015. http://www.xconomy.com/seattle/2009/08/06/stephen

-friend-leaving-high-powered-merck-gig-lights-the-fire-for-open-source-biology-movement/.

Titmuss, Richard. *The Gift Relationship: From Human Blood to Social Policy*. Edited by Ann Oakley and John Ashton. London: Ashton, 1997.

Tkacz, Nathaniel. *Wikipedia and the Politics of Openness*. Chicago: University of Chicago Press, 2015.

Toner, Bernadette. "Harvard's Church Calls for 'Open Source,' Non-Anonymous Personal Genome Project." *Genome Web*, November 1, 2005. Accessed September 12, 2014. http://www.genomeweb.com/harvard-s-church-calls-open-source-non-anonymous-personal-genome-project.

Transcript, House of Representatives, Subcommittee on Oversight and Investigations, Committee on Energy and Commerce: Hearing. *Direct to Consumer Genetic Testing and the Consequences to the Public Health*. Washington, DC, July 22, 2010.

Turner, Fred. *From Counterculture to Cyberculture*. Chicago: University of Chicago Press, 2006.

Tuskegee University. "The Tuskegee University National Center for Bioethics in Research and Health Care: Continuation Application." Tuskegee, AL: Tuskegee University Press, 2000.

Tutton, Richard. "Person, Property, and Gift: Exploring Lauguages of Tissue Donation to Biomedical Research." In *Genetic Database: Socio-Ethcial Issues in the Collection and Use of DNA*, edited by Richard Tutton and Oonagh Corrigan, 17–38. New York: Routledge, 2004.

23andMe. September 15, 2014. "TIME to Thank Our Friends." *23andMe Blog*, October 31, 2008. http://blog.23andme.com/news/time-to-thank-our-friends/.

———. September 22, 2014. "Google Co-Founder Blogs about 23andMe Data, Parkinson's Risk." *23andMe Blog*, September 18, 2008. http://blog.23andme.com/news/google-co-founder-blogs-about-23andme-data-parkinsons-risk/.

UNESCO. *What Is Race?* Paris, 1952.

"Using DNA Matching to Crack Down on Dog Droppings." *NPR.org*. Last modified March 26, 2016. http://www.npr.org/2016/03/26/471958024/using-dna-matching-to-crack-down-on-dog-droppings.

Van Dijck, José van. *The Culture of Connectivity: A Critical History of Social Media*. Oxford: Oxford University Press, 2013.

Vaugham, Adam. "How Viral Cat Videos Are Warming the Planet." *The Guardian*, September 25, 2015. https://www.theguardian.com/environment/2015/sep/25/server-data-centre-emissions-air-travel-web-google-facebook-greenhouse-gas.

Venter, J. Craig. *A Life Decoded: My Genome, My Life*. New York: Viking, 2007.

———. "We Have Learned Nothing from the Genome." By Spiegel. July 29, 2010.

Vermeir, Koen, and Daniel Margocsy. "States of Secrecy: An Introduction." *British Society for the History of Science* 45, no. 2 (2012): 1–12.

Wade, Nicholas. "A Decade Later, Genetic Map Yields Few New Cures." *New York Times*, June 12, 2010. http://www.nytimes.com/2010/06/13/health/research/13genome.html?pagewanted=all&_r=0.

———. "First Gene to Be Linked with High Intelligence Is Reported Found." *New York Times*, May 4, 1998. http://www.nytimes.com/1998/05/14/us/first-gene-to-be-linked-with-high-intelligence-is-reported-found.html.

———. "Genes Show Limited Value in Predicting Disease." *New York Times*, April 15, 2009. http://www.nytimes.com/2009/04/16/health/research/16gene.html.

———. "Researchers Say Intelligence and Diseases May Be Linked in Ashkenazic Genes." *New York Times*, June 3, 2005. http://www.nytimes.com/2005/06/03/science/researchers-say-intelligence-and-diseases-may-be-linked-in.html.

———. "Scientists Complete Rough Draft of Human Genome." *New York Times*, June 26, 2000. https://partners.nytimes.com/library/national/science/062600sci-human-genome.html.

Wadman, Meredith. "Gene-Testing Firms Face Legal Battle." *Nature* 453 (2008): 1148–49.

Wailoo, Keith. *Dying in the City of the Blues: Sickle Cell Anemia and the Politics of Race and Health*. Chapel Hill: University of North Carolina Press, 2001.

Wakeford, Tom, and Fiona Hale. *Generation Scotland: Towards Participatory Models of Consultation*. Newcastle: Policy Ethics and Life Sciences Research Institute (PEALS): University of Newcastle, 2004.

Wald, Priscilla. "Cells, Genes, and Stories: Hela's Journey from Labs to Literature." In *Genetics and the Unsettled Past: The Collision of Race, DNA and History*, edited by A. N. Keith Wailoo and Catherine Lee. Newark, NJ: Rutgers University Press, 2012.

Waldby, Catherine, and Robert Mitchell. *Tissue Economies: Blood, Organ, and Cell Lines in Late Capitalism*. Durham, NC: Duke University Press, 2005.

Wall, Jeffrey D., and Jonathan K. Pritchard. "Haplotype Blocks and Linkage Disequilibrium in the Human Genome." *Nature Reviews Genetics* 4, no. 8 (2003): 587–97.

Washington, Booker T. *Up from Slavery*. Mineola, NY: Dover Publications, 1995.

Wellcome, December 18, 2014. Accessed January 22, 2015. http://www.wellcome.ac.uk/News/Media-office/Press-releases/2014/WTP055974.htm.

Wells, Spencer. *The Journey of Man: A Genetic Odyssey*. Princeton, NJ: Princeton University Press, 2002.

Wertz, Dorothy C. "Genetic Discrimination—an Overblown Fear?" *Nature Reviews Genetics* 3, no. 7 (2002): 496.

Whitfield, Nicholas. "Who Is My Stranger? Origins of the Gift in Wartime London, 1939–45." *Journal of the Royal Anthropological Institute* (May 2013): S95–S117.

Williams, Alex. "On the Tip of Creative Tongues: Curate." *New York Times*, October 2, 2009. http://www.nytimes.com/2009/10/04/fashion/04curate.html.

Williams, Timothy, and Alan Blinder. "Baton Rouge Attack Deepens Anguish for Police: 'We've Seen Nothing Like This.'" *New York Times*, July 17, 2016. http://www.nytimes.com/2016/07/18/us/baton-rouge-attack-deepens-anguish-for-police-weve-seen-nothing-like-this.html.

Winickoff, David. "Genome and Nation: Iceland's Health Sector Database and Its Legacy." *Innovations: Technology, Governance, Globalization* 1, no. 2 (2006): 80–105.

Winterhalter, Benjamin. "A Genetic 'Minority Report': How Corporate DNA Testing Could Put Us at Risk." *Salon*, January 26, 2014. http://www.salon.com/2014/01/26/a_genetic_minority_report_how_corporate_dna_testing_could_put_us_at_risk/.

Wojcicki, Anne. "Research Participants Have a Right to Their Own Genetic Data." *23andMe Blog*, November 4, 2009. Accessed September 15, 2014. http://blog.23andme.com/23andme-research/let-research-participants-access-their-genomes/.

———. "23andMe Provides an Update Regarding FDA's Review." *23andMe Blog*, December 5, 2013. Accessed December 22, 2015. http://mediacenter.23andme.com/?p=1978.

"Your Data Are Not a Product." Editorial. *Nature Genetics* 44, no. 4 (2012): 357–57.

Zoëga, Tómas, and Bogi Andersen. "The Icelandic Health Sector Database: Decode and the 'New' Ethics for Genetic Research." In *Who Owns Our Genes? Proceedings of an International Conference*, 33–64. Copenhagen: Nordic Committee on Bioethics: Nordic Council of Ministers, 1999.

INDEX

Abbott Laboratories, 54
Account of Oneself, An (Butler), 84
Act on Health Sector Database, 97–98
Africa, 18, 116; Out-of-Africa theory, 47, 49
African Americans, 47–48, 53–54, 70, 79, 158, 160–61, 173, 190, 193, 229n11, 230n17, 230n23; "bad blood," treatments for, 55; DNA sampling of, 50, 61–62, 66; genetic testing, 230–31n26, 236n93; heart disease, study of, 61, 68–69; human genome research, 57–60, 62, 64–66, 70, 195; in human genomics, 51; and hypertension, 49, 235n85; Jackson Heart Study, 237n111
African Burial Ground (ABG), 230–31n26
Against Race (Gilroy), 48
Agencourt Personal Genomics, 157
Agent Orange, 25
Alabama, 59, 234n76, 235n89
Alameda County, 193–94
Alameda County Network Program for Reducing Cancer Disparities, 193
Alder Hey Children's Hospital, 100
Althing, 97–99, 118
Amazon, 19, 181
American Association for the Advancement of Science on the Human Genome Project, 65
American Red Cross, 200
American Revolution, 159
American Society of Human Genetics, 53
Angrist, Misha, 149, 153–54, 156, 158, 162, 166, 174, 262n12
Aniag, Jaennika, 267n79
antigens, 230n18
antiracial genomics, 9
antiracism, 48, 56–57
ApoB study, 61–62, 68–69
Appiah, Anthony Kwame, 116
Apple, 1, 120, 128
Applied Biosystems, Inc. (ABI), 16–18, 31–33, 36, 38, 42, 149, 153, 157, 176, 256n15
Aquinas, Thomas, 262n14
Arab Spring, 218n47
Arendt, Hannah, 1, 16, 18–19, 22, 117, 122, 145, 169, 173, 194, 199, 228n89, 260n85, 260n91, 262n13, 269n2; and automation, 222n113; on evil, 207; "inter-est," notion of, 21, 184, 192, 265n50; public realm, publicity of, 175; scientific thought, 162; social economy, 162–63; on speech, 15, 187, 192, 221n92; stories, as remembering, 192–93; on truth, 167
Argentina, Dirty Wars in, 22
Aristotle, 187, 262n14
Arrington, David, 179
arts of collective judgment, 184–85
Ashburner, Michael, 29
Asia, 85, 116
Assange, Julian, 156
Athena Breast Health Network, 177, 188; Athena trial, 183
atomic age, 25
Avey, Linda, 7, 121, 123–25, 130, 132–33, 143, 199, 250n15

Backdoor to Eugenics (Duster), 51–52
Bacon, Francis, 225n29
Ball, Madeleine, 154
Bangham, Jenny, 265–66n63
Barnett, Ross, 237n111

Baudrillard, Jean, 43
Bayer, 170, 178
BBC, Genomics Initiative, 265–66n63
Beeson, Diane, 236n93
Behnke, Hans, 163
Beijing (China), 110
Beijing Genomics Institute (BGI), 151
Bell Curve, The (Herrnstein and Murray), 6, 53
Bell Labs, 129
Belmont Report, 58, 195–96
Benioff, Mark, 177–78
Benjamin, Ruha, 264n38, 266n73
Bentham, Jeremy, 227n71
Berlin (Germany), 203–4, 206, 208
Bermuda, 28, 33–35, 226n40, 226–27n53
Bermuda principles, 29, 37, 156, 185, 224n12; of open access, 34, 38
Between Past and Future (Arendt), 145
Biden, Joe, 176
big data, 12–13, 183, 198, 200
Big Science, 30, 143
biobank collections, 193; lack of diversity, 158
biocapitalism, 128
biocolonialism, 19, 70
bioconstitutional citizenship, 75
bioconstitutionalism, 14, 238n22
bioethics, 56, 75, 85, 89, 91, 236n104
bioinformatics capitalism, 181
biological citizenship, 89
biological colonialism, 46
biological racism, 46
biomedical research, 8, 55–56, 71, 87, 90, 93, 98, 102, 148, 188
biomedical studies, 87; women, underrepresentation in, 55–56
biomedicine, 13, 20, 182, 185, 202; exploitation by, 9; and genomics, 180; informatic approach to, 176–77; social contract with, 200–201
biosociality, 238n22
Black Belt, 59–61, 234n76, 237n111; DNA samples, collecting of, 60; and genomics, 62, 64
Blair, Tony, 37, 229n99
Blakey, Michael, 230–31n26
Bliss, Catherine, 217n34
blood donations, 247n85, 249n107; as civic duty, 200
Bloom's syndrome, 141
Bobe, Jason, 150–54, 157–60, 257n28
body: informed consent, 117; as intimate space, 117
border walls, 216n14
Boston, MA, 110
Botstein, David, 57–58
Brand, Stewart, 44, 128
Brazil, 70. *See also* BRIC
Brenner, Sydney, 31, 37
BRIC, 70. *See also* Brazil; China; India; Russia
Brin, Sergey, 123–24, 126, 130, 199, 250n14
Britain, 34, 81. *See also* England; Scotland; United Kingdom
Broad Institute, 38
Brookings Institution, 6, 170
Brown, Wendy, 174
Busby, Helen, 247n85
Bush, Vannevar, 30
Butler, Judith, 84, 116

California, 112–13, 119, 132, 177, 264n38
California Department of Public Health, 122
Cambridge (England), 95, 110, 115
Canada, 73, 81
Cancer Genome Atlas, The (TCGA), 193
capitalist economy, 32
Casey, Lisa, 121
Cavalli-Sforza, Luca, 46, 58
Celera, 29, 31, 38, 129, 131, 183, 215n11, 236–37n106, 256n15
Center for the Study of Public Genomics, 148–49

Center for Transdisciplinary ELSI Research in Translational Genomics, 193
Centers for Disease Control (CDC), 56
Centre d'Étude du Polymorphisme Humain (CEPH), 50, 258n49
Chadwick, Ruth, 151, 166
China, 70, 73, 79, 81–82, 111. *See also* BRIC
Christensen (congressman), 135
Church, George, 7, 44–45, 130, 148–49, 153–54, 156, 158, 161, 166, 168, 256n15, 256n22, 257n32, 261n111; open technology, belief in, 151–52; and Polonator, 157; sequencing technology, as main goal of, 160
citizenship, 207, 251n39, 259n66; blood donation, 200–201, 247n85; flesh and blood, linkage to, 200; tissue donation, 200
civic institutions, 189
Civil Rights Act (1964), 79
civil rights movement, 79
Clinical Laboratory Improvement Amendments (CLIA) (1988), 150
Clinton, Bill, 2, 8, 24, 37, 41, 56, 59, 66, 99, 215n7, 215n8, 229n99
Colbert, Stephen, 160
Cold Spring Harbor Laboratory, 224n12
Cold War, 70, 72, 203
Collins, Francis, 8, 28–29, 33, 41, 47–48, 53, 66–67, 72, 84, 188, 215n11, 236–37n106; SNP map, 54
colonialism, 47, 74
ColorOfChange.org, 171
Common Disease-Common Variant (CD-CV) hypothesis, 1, 92, 109
Common Rule, informed consent, 181–82
Common Thread, The (Sulston and Ferry), 25, 31–33, 35–36
communicative capitalism, 128
Communities of Color and Genetics Policy Project (CCGPP), 62, 64–65, 173, 195, 197
Complete Genomics, 151
computers: automation, and reflective thought, 16; computer age, 16; and knowledge, 16; personal computers, 15
concentration camps, 203
Congressional Black Caucus, 56
Congressional Caucus for Women's Issues, 55–56
Congressional Oversight Committee, 138
Cook-Deegan, Robert, 148–49
Cooper, Richard, 235n85
Coriell Institute, 75, 77–78, 156–58, 178, 239n29, 263n27
cosmopolitan ethics, cosmopolitanism, and sharing, 116
Council for Advancement and Support of Education, 179
Council for International Organizations of Medical Sciences (CIOMS), 90
Crawford, James M., 119
Creative Commons, 154
Crick, Francis, 31, 149
curation, 139–40, 142
cybernetics, and genomics, 40

Darwin, Charles, Out-of-Africa theory, 49
Daston, Lorraine, 253n73
data smog, 43
Dean, Jodi, 40–41, 128, 181
deCODE Genetics, Inc., 94, 97, 100, 103–4, 107, 137, 242–43n18; commodification, concerns over, 98; controversy over, 98–99
deCODEme, 137
Denmark, 97
Denver, CO, 74, 89, 240–41n70
Department of Energy, 26
Descent of Man (Darwin), 49
Desmond-Hellman, Susan, 10–11, 145, 147–48, 179, 200–201
Diamond v. Chakrabarty, 30
digerati, 120, 132–34, 139

digital age, 44
digital automation, 42
digital economy, 42
digital utopianism, 130
diversity: biobank collections, lack of, 158; genetic diversity, 46; and genomes, 51; human genetic diversity, 54; human genomes, 49; mapping diversity of, 51; Personal Genomic Project (PGP), lack of, 158–61
DNA, 15, 23, 28, 42, 45–46, 48–49, 73–74, 76, 80, 97, 103–7, 109, 117–18, 120, 133, 136, 145, 149, 153–55, 173–74, 178, 182, 186, 189, 201, 207, 269n11; African Americans, sampling of, 50, 61–62, 66; and cloning, 165–66; as commodity, 98–99; democratizing of, 125; digitization of, 176; for genomic research, 100; and genotyping, 124; home testing, 121; identifying information, 131; management of, 179; as national treasure, 108; phenotype data, 126–27; sequencing machines, 11, 151–52; sequencing of, 13, 17–18, 22, 26–27; sharing of, 12, 147–48, 164; SNP map, 54; sovereign people, and self-government, 95; spit parties, 132
DNA Polymorphism Discovery Resource, 232–33n56
Do-It-Yourself (DIY) movement, 128
Do-It-Yourself Biology (DIYBio) movement, 152; Continental Congress of North America, 159
Dreamforce, 177, 179
Drosophila genetics, 29, 31, 224n11
Du Bois, W. E. B., 48–49
Duffy Antigen/Chemokine Receptor (DARC), 59–62, 234n77
Duke University, 31; Institute for Genome Sciences and Policy, 148, 224n12
Dunston, Georgia, 49–50, 57–59, 161, 230–31n26
Duster, Troy, 51–53, 58, 231n38, 236n93
Dylan, Bob, 28
Dyson, Esther, 126, 159

East Germany, 260n89. *See also* Germany
Economic and Social Science Research Council (ESRC), 244n34; Genomics Forum, 242n9
Edinburgh (Scotland), 23, 94–96, 196, 207
Eli Lilly, 161
Elizabeth II, 33
ENCODE, 198
England, 32, 95, 97, 111–12, 242n7, 247n83. *See also* Britain; United Kingdom
Enlightenment, 8, 26, 101, 118, 241n4, 262n14
Episcopal Church, 64–65
Epstein, Steven, 55
Erlich, Yaniv, 155–56, 258n49
Esserman, Laura, 177, 179, 188
Estonia, 97
Estonian Genome Project, 94, 97
ethics, 19, 24, 57, 80, 91, 96, 112–14, 148, 155, 160, 173–74; cosmopolitan ethics, 99, 116; good ethics, and good science, 27; and justice, 115, 166; knowledge, problem of, 115
Etzioni, Amitai, 10
eugenics, 5, 52, 207
Europe, 29, 83, 85, 98, 116, 164, 244n41; xenophobia in, 206
European Union (EU), 4, 13
Evans, Jim, 135–36

Facebook, 13, 120, 129, 164–65, 169, 181, 262n2
fascism, 29–30
Federal Drug Administration (FDA), 122, 141, 255n101
Ferry, Georgina, 25
Fletcher, John, 56
Foner, Eric, 259n66
Food and Drug Administration (FDA), 121
Fortun, Mike, 223n3

Foster, Henry W., Jr., 56
Foucault, Michel, 260n91
Fourth Amendment, 11
Framingham, MA, 116
Framingham Heart Study, 235n86
Friend, Stephen, 154
From Counterculture to Cyberculture (Turner), 128
"From Genetic Privacy to Open Consent" (Lunshof et al.), 151
Fullerton, Malia, 258n49

Gandhi, Indira, 88
GATTACA (film), 220n82
GenBank, 33–34
gene mapping, 31; moral economy of, 32
Genentech, 10
Generation Scotland (GS), 94–96, 100, 102, 110, 123, 173, 176, 179–80, 184, 186, 192, 242n10, 243n29, 244n34, 244n39, 245n42, 245n45, 246n62, 247n82, 247n83, 248n98; benefit sharing, 105–7; database, access to, 104–5, 117; DNA samples, staying in Scotland, 108–9, 111–15; empiricist model, 104; focus groups, 104; human altruism, 116–17; informed consent, 118; major pharma investment, evaporation of, 107; Management, Access, and Policy Publication (MAPP), 105–6; public sovereignty, 108; Scientific Committee, 104; and value, 114; will of the people, as embodiment of, 101, 103–5, 108
Generation Scotland Legal and Ethical Aspects Report, 108
Genetic Information Nondiscrimination Act (2008) (GINA), 257n43
genetics, 7, 22, 51, 124; genetic sequences, 42, 149, 150; genetic studies, 155–56; and IQ, 6; lay public, accessibility to, 125; and race, 58; representation, question of, 81–82; and trust, 5
Genetics Core of the Wellcome Trust Clinical Research Facility, 109
Genetics Healthcare Initiative, 102
genomes, 2, 14–15, 18, 27, 39–40, 51, 54–55, 93–94, 96, 117, 119, 123, 145, 169; as agent of change, 99; bioconstitutional issues, 75; common humanity, demonstrating of, 66–67, 70; constitutional of, 92; cosmopolitan ethic of care, 116; democratic sovereignty, 115; differences between, 6, 47; and diversity, 49; draft of, 99; and haplotypes, 71; human genome maps, 78, 88, 92; as "in-formation," 118; interpreting of, 198; justice, sense of, 10; as meaningful, 20; national borders, 115; of national importance, 118; as national resource, 97; next generation sequencing methods, 109–10; rights, issue of, 76; sequencing machines, 153; sequencing of, 28, 131, 168, 174–75, 207, 215n11; sharing of, 10–11, 31–32, 148; social media strategies, 126; and trust, 5; value, drop in, 115–16
"Genomes for All" (Church), 149, 152
genome-wide association studies (GWAS), 123–24
genomic data: phenotype data, 126; sharing of, 179
genomic information, 40, 178; as fundamental right, 134; and knowledge, 39; and meaning, 41, 45–46; value of, 44
genomic liberalism, 7, 9
Genomic Research in African American Pedigrees (G-RAP), 50–54, 58, 67
genomics, 1, 4, 6–7, 10, 14–15, 20–21, 23–24, 30–31, 34–35, 42, 46, 48, 59, 87, 94–95, 97–98, 159, 162, 169, 171, 173–74, 199–200, 220n87, 223n3, 225n31; access to all, 152, 186; African Americans, 51, 53–54, 190, 195; and bioinformatics, 182–83; and biomedicine, 180; civic spaces, 190; colonization, fear of, 142; "copyleft" agreement, 38; and cybernetics, 40; data, as public, 176; data, sharing of, 185, 188–89; data management,

genomics (*cont.*)
124; decoding, as goal of, 41; democratization of, 91, 124, 133; digital computers, 198; genomics research, 47, 68, 160, 182, 189, 197; as high-prestige science, 74; human diversity, 51; and inclusion, 22; as international effort, 70–71; justice, claims to, 18–19; and knowledge, 197–98; and liberalism, 40; management, shift to, 178; and meaning, 44; mob rule, fear of, 135; open consent, 158; and plurality, 187; precision medicine, 179; and privacy, 156, 168; as public good, 188–89; and race, 49, 55, 58; race, social concepts of, 19; and recognition, 84–85; representation, question of, 81–82; social networking platforms, 120; society, computerization of, 16; technical elite, favoring of, 8, 13; value of, 96, 122
Genomics Policy and Research Forum, 96
Genset of Paris, 54
GenSpace, 153
Germany, 97, 163, 204, 207. *See also* East Germany
Ghostly Matters (Gordon), 1
Gift Relationship, The (Titmuss), 200
Gilroy, Paul, 48–49, 116
Gingrey, Phil, 135
Glasgow (Scotland), 171, 182
GlaxoSmithKline, 178
Gleick, James, 44
global capitalism, 181
globalization, 91, 202
Goetz, Thomas, 250n14
good science, 261n110; good society, 184
Google, 1, 13, 120, 122–24, 130, 141, 143, 155, 174, 199, 250n14; Google buses, 14
Gordon, Avery F., 1
Graham, Bettie, 53
Greely, Henry T., 98
Greenberg, Judith, 75
Greenlining Institute, 264n38
Guthrie, Woody, 28

Hacker's conference, 44
Haley, Alex, 50
Haraway, Donna, 223n129, 262n8, 264n39, 266n76
Harris, Parker, 177–78
Harris-Wai, Julie, 193–94, 267n79
Harvard University, 157, 258n47; Harvard Data Privacy Lab, 155–56, 165; Harvard Law School, Bill of Rights Project, 165; Personal Genome Project (PGP), 130, 148–49
Hawley, R. Scott, 29
Hayden, Cori, 167
Health Insurance Portability and Accountability Act (1996) (HIPAA), 150
HeLa cells, 9, 193
Helix, 1, 215n1
Herrnstein, Richard J., 6
Higginbotham, Evelyn Brooks, 48
Hilgartner, Stephen, 30
Hiroshima (Japan), 25, 44, 170
Hitler, Adolf, 203
HIV/AIDS, 145
Hoffmann-La Roche, 98, 242–43n18
Holocaust, 15–16
hormone replaced therapy (HRT), 102–3
Houston, TX, 72, 89, 192, 207, 240–41n70; Indian American community in, 88
Howard University, 54–55; National Human Genome Center, 66
Hubbard, Tim, 38
Human Condition, The (Arendt), 1, 15, 19, 187, 228n89, 269n2
Human Genome Diversity Project (HGDP), 22–23, 46, 65, 66, 70–71, 73–75, 78, 85, 92, 98; antiracist potential of, 52; colonialism, charges of, 47; indigenous populations, DNA samples from, 5–6; scien-

tific racism, fear of, 52; Vampire Project, dubbing of, 5, 52
Human Genome Organization (HUGO), 226n40
Human Genome Project (HGP), 2, 6, 8, 11–13, 17, 19–20, 22, 25–27, 30–31, 33–35, 37–38, 43–45, 47, 50–53, 57–59, 65–68, 75, 95–96, 107, 109–10, 131, 143, 148–52, 157, 160–61, 167–68, 185, 187–88, 197–98, 224n12, 256n15, 257n32; freedom of information, 28–29; human condition, threatening of, 18; public science, defense of, 28
Human Genome Research Institute (NHGRI), 6
Human Genome Sciences (HGS), 37, 43
human leukocyte antigen (HLA), 49, 161
human sequencing, 1–2, 4, 6–7, 12–13, 16–17, 20, 24, 31, 33, 36–41, 47–48, 58–59, 178, 185; data-sharing agreement, 28–29; expressed sequence tags (ESTs), 28; private property, 32; as public good, 188; sequencing machines, 18, 35, 156–58, 160
Human Sequencing and Mapping Index, 226n40
Hutchinson, Richard, 237n111

Ibadan (Nigeria), 72, 85, 171, 182, 192, 196
Iceland, 95, 98, 100, 108, 118, 173; deCODE in, 94, 97–99, 107
Icelandic Health Database, 140
Icelandic Medical Association, 98
Illumina, 18, 170, 262n12
Immortal Life of Henrietta Lacks, The (Skloot), 8–9, 193
Immunogenetics Laboratory, 50
Inclusion (Epstein), 55
India, 70, 73. *See also* BRIC
indigenous populations, 52, 71, 74, 85; DNA sampling, 5, 46, 182. *See also* Native Americans, genomic research
informatic capitalism, 27, 37, 175, 184–85, 202; and openness, 38–39
informatics, 16, 122; biology, constitution of, 199–200; and genomics, 127; human freedom, 128; and mathematics, 15
information, 181, 195; encoding of, 40; as free, 44; information overload, 43; signals, transmission of, 40–41
Information, The (Gleick), 44
informed consent, 67–68, 77, 85–88, 90, 117–18, 155, 163–64, 259n69; Common Rule, 181–82; educated population, 98
Institute of Genome Research (TIGR), 42–43
International Commission for Protection against Environmental Mutagens and Carcinogens, 26
International Haplotype Map Project (HapMap), 6–7, 70, 73–74, 77, 91, 93, 96, 131, 178–80, 184–86, 188, 190, 217n34, 238n9, 240–41n70, 263n27; bioconstitutional questions of, 92; blood donation, 88–89; and communities, 79, 82–85; community engagement policy, 78, 82, 85–88, 92; as contentious, 80; focus groups, 85–86; inclusion, practices of, 71–72; informed consent, 85–88, 90; International Hapolyte Map, 89; as international science project, 88; naming of self, 84; precision, commitment to, 80, 81–82, 85; as race map, accusations of, 80; and recognition, 84–85; and representation, 89, 92
Internet Archive, Open Library project, 147
Italy, 74

Jackson, Fatimah, 50, 57–59, 61–62
Jackson-Morgan, Joi, 171
Jameson, Fredric, 43
Japan, 73, 79, 82, 85
Jasanoff, Sheila, 14, 216n18, 268n105; bioconstitutionalism, as term, 238n22
Jenkins, Henry, 251n39

Johns, Adrian, 34
Johns Hopkins University, 9, 147
judgment, 26, 33, 39, 122, 186, 194, 196, 199–200; arts of collective judgment, 184; and speech, 187
justice, 8–11, 22–24, 26, 44, 56, 67, 70, 72, 85, 93, 106, 115, 134, 147, 159, 166, 168, 174, 183–85, 194, 197; collective actions, 195–96; and democracy, 133; and equity, 103; and genomics, 18–19; and knowledge, 25, 27, 48, 113, 121–23, 162, 173, 202; and science, 160, 196, 202; and truth, 16, 18, 148, 167, 207
Justice (Sandel), 10

Kant, Immanuel, 187
Kennedy, John F., 237n111
Kent, Jim, 29
King, Mary-Claire, 22, 98
King, Patricia, 46, 58, 67, 195
Knight Foundation, 154
knowledge, 9, 18–19, 24, 26, 29–30, 35, 59, 67, 75, 97, 116, 131, 136, 141, 161, 168, 184, 200; and ethics, 96, 115; genomic information, 39, 197–98; and genomics, 197–98; as informational commodity, 37; and justice, 25, 27, 48, 72, 113, 121–23, 162, 173–74, 202; knowledge production, 31, 34, 140, 143; as meaningful, 2, 13, 199; and openness, 34; and politics, 15–16; sharing of, 147
Kohler, Robert, 31–33, 224n11, 226n42, 226n44
Kruglyak, Leonid, 234n77
Kutz, Gregory, 135
Kuwait, 269n11

Laboratory of Molecular Biology (LMB), 31, 35, 37
Labrador, 106
Lacks, Henrietta, 8–10, 147
Lander, Eric, 36, 38, 50
Language of God, The (Collins), 41
Latour, Bruno: matters of concern, 20; technoscience, as term, 225n31
Leon, David, 244n41
Lewontin, Richard, 98
liberalism, 7, 11; and genomics, 40
Life and Times of Lilly Mendel, The (blogsite), 125
Life Decoded, A (Venter), 41
Life Technologies, 254n94
Locke, John, 11, 135
London (England), 94–95
Lu, Lina, 31
Lunshof, Jeantine, 151, 166
Lyotard, Jean-François, 15–16, 18–19, 37, 143, 199

Macon County, 60, 67, 72
Making of the English Working Class, The (Thompson), 32
Mardis, Elaine, 12
Martin, Paul, 247n85
"Mathematical Theory of Information, A" (Shannon), 40
Mayo Clinic, 124
Merck, 154, 161, 178, 260n79
Meredith, James, 237n111
Merton, Robert, 29–30, 161
Mexico, 4, 97, 115, 216n14
Mexico City (Mexico), 23, 207
Meyer, Michelle N., 165–66, 186
Michael J. Fox Foundation, 124
Mississippi, 237n111
Mitchell, Rob, 201
modernity, 190
Moore, Gordon, 17
Moore's Law, 44, 138
moral economy, 32–33, 38, 129, 201; as term, 224n11, 226n42, 226n44
Morgan, Thomas Hunt, 31, 33
Mormons, 50
Morris, Andrew, 101

Murdoch, Wendi, 132
Murray, Charles, 6
mutations, 26, 113–14, 248n98
Myers, Richard, 57
Myriad Genetics, 28, 53

Nagasaki (Japan), 25, 44, 170
National Academy of Sciences, 264n35
National Association for the Advancement of Colored People (NAACP), 79
National Cancer Institute (NCI), 140, 185, 193
National Cancer Moonshot Initiative, 176
National Center for Bioethics in Research and Health Care, 62
National Center for Human Genome Research (NCHGR), 54; Working Group on Ethical, Legal and Social Issues, 52–53. *See also* National Human Genome Research Institute (NHGRI)
National Commission for the Protection of Human Subjects of Biomedical and Behavioral Research, Belmont Report, 67
National Health and Nutrition Examination Survey (NHANES), 232–33n56
National Health Service (NHS), 105–6, 200, 253n68
National Human Genome Research Institute (NHGRI), 6, 46–47, 54, 57, 59, 61, 70, 80, 82–87, 89–90, 131, 155, 192, 217n32, 229n1, 232–33n56, 239n29; community engagement, 77–78; human genetic variation, map of, 72–73; informed consent, 77; large populations, sampling of, 71, 74; two-pronged approach of, 74. *See also* National Center for Human Genome Research (NCHGR)
National Institute of General Medical Sciences (NIGMS), 78–79; cell repository, 76–77; community advisory group, 77; consultation, call for, 76–77; diverse populations, cell lines of, 75; and privacy, 75; and rights, 77
National Institute of Genomic Medicine, 97
National Institutes of Health (NIH), 6, 8, 27, 33, 42, 52–54, 56, 59, 61, 72–73, 75–76, 78–80, 90–91, 113, 143, 153, 155–56, 159, 161–62, 173, 179, 186, 188, 192, 196, 201, 215n11, 217n32, 231n29, 232n46, 255n98, 257n43; Advanced Sequencing Technology Award, 17; health infrastructures, inequities between, 93; and privacy, 168; Women's Health Initiative, 102
National Institutes of Health Revitalization Act (1993), 55–56
National Science Foundation (NSF), 196, 222n124
Native Americans, genomic research, 70, 223n127. *See also* indigenous populations
Navigenics, 132, 137–40, 142, 254n77, 254n94
Nazis, 163
Nelson, Alondra, 230n12
Newfoundland, 106
New Mission, 173
New Rochelle, NY, 74, 89, 240–41n70
Newton, Isaac, 153
New York City, 200–201
Nickerson, Deborah, 61, 237n111
Nigeria, 79, 81, 182
9/11 attacks, 200–201
Nixon, Rob, slow violence as term, 223n3
North America, 116
Norway, 97
"Notes on Science and Democracy" (Merton), 29–30
Nunberg, Geoffrey, 43–44

Obama, Barack, 33, 175–76, 188
objectivity, 139, 253n73
Occupy Wall Street, 190
Office of Technology Assessment, 197

Ole Miss. *See* University of Mississippi
Olson, Maynard, 57–61
On Violence (Arendt), 169
Open Human Network, 154
Open Humans, 7, 154, 257n28
openness, 183, 186–87, 225n29; health records, digitizing of, 175–76; informatic capitalism, 38–39; and judgment, 187; of medical information, 175; open data sharing, 154; open genomes, and open humans, 155; open science, 153; and transparency, 175
Oppenheimer, J. Robert, 26
Oxford (England), 95

Parkinson's disease, 123–24, 126
Parnell's mustached bats, 20
Participation Information Leaflet (GS), 108–9
Pathway Genomics, 134
Patient Information Leaflet (GS), 111
Pauling, Linus, 236n101
Payton, Benjamin F., 68, 233n67
Pepco, 183
Perlegen, 123–24, 126
Personal Genome Project (PGP), 7, 30, 31, 149–50, 153, 157, 165, 176, 183, 186, 257n43, 258n47, 258n49; consent, and veracity, 166; demographics of, 158; diversity, questions about, 158–61; Do-It-Yourself spirit of, 257n18; free access, 152; Genes, Environment, and Traits (GET) conference, 159; informed consent form, 155, 163–64, 259n69; justice, claims to, 159; just science, 161; open access, 158, 167–68; open consent policy, 151; open humans, study of, 154; and privacy, 148, 156, 162, 166–67; privacy rights, as impeding openness, 161; public genomics research, goal of, 160–61
PersonalGenomes.org, 154, 166
personal genomics (PG), 120, 127, 130, 155; biosocial formation, 123; concerns over, 134–36; democratization of, 121, 133; and knowledge, 121–22; mainstream, entering into, 134; mob rule, fear of, 135; science, corporate corruption of, 140–41
personal genomics companies, 121, 123, 134, 199; accessible information, 143–44; and curation, 139–40; early adopter approach, 136; industry standards, creating of, 136–38; personal genomics services, 137–38; persons, standards of, 143; profits, need for, 142; science, corporate corruption of, 140; as untrustworthy, 141
Personalized Medicine Coalition (report), 143
Personal Medicine Coalition (PMC), 136–37
Pfizer, 107, 170
Piracy (Johns), 34
poiesis, 174
"Policy for the Responsible Collection, Storage, and Research Use of Samples from Identified Populations for the NIGMS Human Genetic Cell Repository," 76
Political Unconscious, The (Jameson), 43
Portable Legal Consent (PLC), 154
Poschardt, Ulf, 204, 206
Poster, Mark, 43
postgenomic condition, 2, 12, 14–15, 19–22, 39, 95, 115–16, 118, 144, 174, 184, 191, 200, 202; concerns of, 185; crisis of value, 99; justice and knowledge, questions of, 72; meaning, problem of, 24, 27, 192; and plurality, 187; thought and life, displacement of, 185
Postmodern Condition, The (Lyotard), 15, 143
post-truth politics, 4
precision medicine, 179; data, sharing of, 180

Precision Medicine Initiative (PMI), 188–89, 197; Common Rule, as crucial to, 181
privacy, 10–11, 21–22, 75, 131, 143, 145, 165–66, 176, 206; and collectivization, 167; data privacy, debate over, 13–14; genetic privacy, 151; and genomes, 156, 168; privacy rights, 161–63; and publicness, 167
public culture, civic institutions, 189–90
public genomics, 149, 160–61, 176–77; and openness, 168
public good, 159, 184; and genomes, 188–89; privacy rights, 163; sharing data, 191
Public Health Service (PHS), 55
public sovereignty, 108, 116, 241n4
public space, 189–90, 196
Puig de la Bellacasa, Maria, 20
Putnam, Robert, 189–90

Rabinow, Paul, biosociality as term, 238n22
race, 5, 46–48, 67; and genetics, 58; and genomes, 19, 49, 55, 58; and genomics, 19, 49, 55, 58
racism, 23, 74
RAND Corporation, 175–76
Rawls, John, 184, 195
Reagan, Ronald, 52
Reddy, Deepa, 88
Regenesis (Church), 151
Reverby, Susan, 56
right of autonomy, 84
Robert Wood Johnson Foundation, 154
Roots (Haley), 50
Rotation (film), 163
Rotimi, Charles, 235n85
Rousseau, Jean-Jacques, 103, 135, 241n4, 244n39
Roy, Arundhati, 191
Royal Society, 225n29, 244n33
Russia, 70. *See also* BRIC

Sage Bionetworks, 154
Salesforce, 177–78, 191
Sandel, Michael, 10–11
San Francisco, CA, 14, 164, 169, 177, 190, 193, 208, 262n2; Bayview-Hunters Point, extreme poverty of, 170–71; Third Street, 170–71, 178, 180, 191–92, 202–4, 207
Sanger, Fred, 109
Sanger Centre, 35–38
Sanger Institute, 33–34, 111–13, 115; Sanger sequencing methods, 109–10
Savio, Mario, 128
Scalia, Antonio, 11
Schloss, Jeff, 155
science: as communal good, 30; corporate corruption of, 121, 140–41; and inequality, 197; and justice, 202; and medicine, 185–86; openness, as beacon of, 30; secrecy, as antithesis of, 30; technology studies, 22
Science and Justice Working Group, 196
Scientific Committee for Generation Scotland, 101
scientific racism, 52, 55, 57
Scotland, 94–95, 100, 103, 108–11, 113, 115–20, 173, 178, 182, 184, 192, 242n7, 243n29, 245n42; devolution in, 102; genomic research, 99; Independence Referendum in, 96; Scottish genomes, value of, 96–97; as "sick man of Europe," 101, 244n41. *See also* Britain; United Kingdom
Scottish Enlightenment, 102–3
Scottish Family Health Study, 115
Shannon, Claude, 40
Shapin, Steven, 225n31
Shuren, Jeff, 134–35
Silicon Valley, 8, 14, 23, 95, 110, 113, 120, 122–23, 128, 130, 147
single nucleotide polymorphism (SNP), 19, 123–24, 137–39, 232–33n56

Skloot, Rebecca, 8–9, 193
slavery, 47, 50–51
Smith, Adam, 32, 245–46n55, 246n56, 262n14; invisible hand metaphor of, 103
Smith, Ed, 57, 59–62, 68
Smith, Gordon H., 121
Smith, Mike, 36
Snowden, Edward, 14, 156
social media, 218n47, 260n91
Springsteen, Bruce, 28
Standing Committee on Human Genetics Research, 106
Stanford University, 111–12, 129, 140
Staudte, Wolfgang, 163, 260n89, 260n91
Stefánsson, Kári, 97–98
Steinberg, Wally, 37
Stevens, Hallam, 36, 199–200
Stewart, John, 38
Sturdy, Steve, 265n46, 268n99
Sulston, John, 25–26, 28–29, 31–33, 35–39, 159–60, 185, 225n33, 226n40, 260n79
Sunder Rajan, Kaushik, 128, 219n61
Sutter Hill Venture, 215n1
Swartz, Aaron, 147
Sweeney, Latanya, 155–56, 165, 258n49
Synapse, 154

Tahrir Square, 190
Teilhard de Chardin, Pierre, 22
Tempelhofer Feld, 203–4, 206–8
Terranova, Tiziana, 40–41; milieu, as term, 228n80
Thatcher, Margaret, 200
Theory of Justice, A (Rawls), 184
Third Street Youth Center and Clinic, 171
"This Land Is Your Land" (Guthrie), 28
Thompson, Charis, 184, 261n110, 264n38
Thompson, E. P., 32–33, 224n11; moral economy, as term, 226n42
Thurston, Robin, 1
Titmuss, Richard, 200
Tocqueville, Alexis de, 135, 189–90
Tokyo (Japan), 78, 85
Translational Medicine Research Collaboration, 103
Trump, Donald, 262–63n16, 263n23
Turner, Fred, 128
Tuskegee, AL, 23, 55, 59, 64–66, 171, 182, 186, 190, 192, 194, 196, 207
Tuskegee Institute, 67
Tuskegee Institutional Review Board (IRB), 68
Tuskegee Syphilis Study, 9, 67–68, 87, 178, 182, 195, 239n34; apology for, 56, 59, 62; scientific racism, as synonymous with, 55, 57
Tuskegee University, 69, 87, 173, 229n11, 233n67; DNA samples, collecting of, 60–62; genome center, 58–59, 68; heart disease, and African American study, 61; Kellogg Conference Center, 56–57
23andMe, 7–8, 120, 122–23, 130–31, 137, 139, 140, 142–43, 173–74, 178–79, 200, 249n8, 252n48, 253n71, 255n101; and biocapitalism, 128; class action lawsuit against, 121; demographic, targeting of, 132; FDA, problems with, 141; Genetics 101, 133; genomic information, as fundamental right, 134; lay public, accessibility to, 125–26, 133; moral economy, as threat to, 129; personal genomics, 127; personal genomics, democratizing of, 133–34; personal technology, as hip piece of, 128; playful culture of, 250n17; as research revolution, 126; social media strategies of, 126–27; spit kits, 124; spit parties, 132; "To Do" box, 127
Two Treatises on Government (Locke), 11

UK Biobank, 104, 106–7
Undoing the Demos (Brown), 174
United Kingdom, 4, 28, 73, 96, 99, 102, 109, 111–13, 115, 118, 164, 166–67, 188, 200,

207, 243n29, 253n68; blood donation, 247n85, 249n107; science and society initiatives in, 100. *See also* Britain; England; Scotland
United Nations (UN), 5
United States, 4, 28, 47, 59, 69–74, 78–79, 81, 86–89, 119, 148–49, 164, 176, 179, 181–82, 188, 195–97, 200–201, 207, 216n14, 235n85, 240–41n70, 259n66; alt-right, rise of, 228n78; historically black colleges and universities (HBCUs) in, 48, 229n11; individualism in, and public life, breakdown of, 189–90; self-identification in, 239n34
University of California, Berkeley, 140
University of California, San Francisco (UCSF), 169, 180, 201, 219n64; Athena trial, 183; Benioff Children's Hospital, 170, 190–92; informed consent, 182; Medical Center, 145–47; MeForYou campaign of, 10, 12, 164–65, 179, 188, 191, 267n82; New Generation Health Center, closing of, 171; precision medicine, promoting of, 179, 183
University of California, Santa Cruz (UCSC), 29, 122–23
University of Mississippi, 68–69; National Heart, Lung, and Blood Institute (NHLBI) at, 237n111
University of Washington (UW), 57, 61–62
U.S. v. American Bell Telephone Co., 30, 161
Utah, 50, 258n49

Varmus, Harold, 185–86
Vaughan v. Phebe, 162
Venter, Craig, 12, 16, 26, 28–29, 32–33, 36–44, 66–67, 131, 134, 136, 178, 183, 187–88, 215n11, 216n12, 224n22, 226n50, 229n99, 236–37n106
Vietnam, 44
Vietnam War, 25
Virginia Tech University, 68
Vorhaus, Daniel, 151, 166

Waldby, Catherine, 201
Walgreens, 134–35
Warburg Pincus, 215n1
Washington, Booker T., 234n76
Waterston, Bob, 36–37
Watson, James, 18, 28, 35, 43, 131, 134, 148–49
Wealth of Nations (Smith), 103, 245–46n55, 246n56
Weinstein, Harvey, 132
Wellcome Trust, 2, 38, 166–67, 186, 196; Human and Vertebrate Analysis and Annotation group, 200
Wells, Ida B., 52
West Berlin (West Germany), 203. *See also* Berlin (Germany)
Whitehead Institute for Biomedical Research, 155–56
Whole Earth Catalog (Brand), 128
Whole Earth network, 128
WikiLeaks, 156
Wilbanks, John, 154
Wojcicki, Anne, 8, 120–21, 124–26, 130, 132–34, 143, 199
Won for All (Ashburner), 29
World Economic Forum, 133
World Medical Association, 98
World War II, 25, 40, 200–201, 203
Wyeth Pharmaceutical Co., 107; and Prempro, 102–3

Yahoo, 13
yeast artificial chromosomes (YACs), 57–58